The Development of Numerical Competence

Animal and Human Models

Edited by

Sarah T. Boysen
Ohio State University
and
Yerkes Regional Primate Research Center
Emory University

E. John Capaldi
Purdue University

LEA LAWRENCE ERLBAUM ASSOCIATES, PUBLISHERS
1993   Hillsdale, New Jersey   Hove and London

Copyright © 1993, by Lawrence Erlbaum Associates, Inc.
All rights reserved. No part of this book may be reproduced in any form, by photostat, microform, retrieval system, or any other means, without the prior written permission of the publisher.

Lawrence Erlbaum Associates, Inc., Publishers
365 Broadway
Hillsdale, New Jersey 07642

Library of Congress Cataloging-in-Publication Data

The Development of numerical competence : animal and human models / edited by Sarah T. Boysen, E. John Capaldi.
    p.   cm. — (Comparative cognition and neuroscience)
    Includes bibliographical references and index.
    ISBN 0-8058-0749-7 (c). — ISBN 0-8058-1231-8 (p)
    1. Number concept in animals. 2. Learning in animals.
I. Boysen, Sarah Till, 1949–   . II. Capaldi, E. John.
III. Series.
QL785.24.D48   1993
156'.3—dc20   92-32928
                                                                                              CIP

Books published by Lawrence Erlbaum Associates are printed on acid-free paper, and their bindings are chosen for strength and durability.

Printed in the United States of America
10 9 8 7 6 5 4 3 2 1

The Development of Numerical Competence

Animal and Human Models

Comparative Cognition and Neuroscience
Thomas G. Bever, David S. Olton, and Herbert L. Roitblat, Series Editors

Antinucci: Cognitive Structures and Development in Nonhuman Primates,

Boysen/Capaldi: The Development of Numerical Competence: Animal and Human Models,

Brown: The Life of the Mind,

Kendrick/Rilling/Denny: Theories of Animal Memory,

Kesner/Olton: Neurobiology of Comparative Cognition,

Nilsson/Archer: Perspectives on Learning and Memory,

Ristau: Cognitive Ethology: The Minds of Other Animals: Essays in Honor of Donald E. Griffin,

Roitblat/Herman/Nachtigall: Language and Communication: Comparative Perspectives

Roitblat/Terrace/Bever: Animal Cognition,

Schulkin: Preoperative Events: Their Effects on Behavior Following Brain Damage,

Schusterman/Thomas/Wood: Dolphin Cognition and Behavior: A Comparative Approach,

Zentall/Galef: Social Learning: Psychological and Biological Perspectives,

Contents

Preface   vii

PART I   EMPIRICAL APPROACHES TO COUNTING (OR NUMERICAL COMPETENCE) IN ANIMALS   1

1   Invisible Counting Animals: A History of Contributions from Comparative Psychology, Ethology, and Learning Theory
Mark Rilling   3

2   Counting in Chimpanzees: Nonhuman Principles and Emergent Properties of Number
Sarah T. Boysen   39

3   Numerosity as a Dimension of Stimulus Control
Werner K. Honig   61

4   Counting by Chimpanzees and Ordinality Judgments by Macaques in Video-Formatted Tasks
Duane M. Rumbaugh and David A. Washburn   87

PART II   COUNTING: CRITERIA AND RELATIONS TO BASIC PROCESSES

5   Numerical Competence in Animals: Life Beyond Clever Hans
Hank Davis   109

6   Numerical Competence in Animals: A Conservative View
Roger K. Thomas and Rosanne B. Lorden   127

7	Do Animals Subitize? *Daniel J. Miller*	**149**
8	Quantitative Relationships Between Timing and Counting *Hilary A. Broadbent, Russell M. Church, Warren H. Meck, and B. Carey Rakitin*	**171**

PART III COUNTING IN HUMANS AND ANIMALS: THEORETICAL PERSPECTIVES **189**

9	Animal Number Abilities: Implications for a Hierarchical Approach to Instrumental Learning *E. John Capaldi*	**191**
10	A Conceptual Framework for the Study of Numerical Estimation and Arithmetic Reasoning in Animals *C. R. Gallistel*	**211**
11	Reflections on Number and Counting *Ernst von Glasersfeld*	**225**
12	Chunking, Familiarity, and Serial Order in Counting *Wayne A. Wickelgren*	**245**
	Author Index	**269**
	Subject Index	**275**

Preface

J. B. Watson, in several of his books, gave a prominent place to animal numerical abilities as a route to understanding complex problem-solving capacities in various species. Some of this early work was of fair quality—some of it was not. In any event, for some time now, numerical capacity in animals has been a neglected topic. This has changed. This volume of contributed chapters represents the renewed interest in the study of numerical capabilities in a variety of animals, including pigeons (Honig, chap. 3), rats (Capaldi, chap. 9; Davis, chap. 5), monkeys (Thomas & Lorden, chap. 6; Rumbaugh, chap. 7), and chimpanzees (Boysen, chap. 2; Rumbaugh & Washburn, chap. 4). A core of the chapters emerged from symposia in recent years at the Midwestern Psychological Association (1986, 1987, 1992), the Southern Society of Philosophy and Psychology (1988), and the American Psychological Association (1988). Other chapters (Gallistel, chap. 10; Miller, chap. 7; von Glasersfeld, chap. 11; Wickelgren, chap. 12) provide unique theoretical perspectives on the counting process, subitizing, number concepts and serial learning, with discussions of other issues related to the cognitive substrates of counting or number-related abilities. Additionally, Broadbent, Rakitin, Church & Meck outline a connectionist model of counting and timing, providing a theoretical overview of their significant studies of the relationship between these two processes.

Studies of number concepts and early attempts to train other species to demonstrate "counting" has an unquestionably colorful history, clearly reflecting the scientific Zeitgist at the time (Rilling, chap. 1). Similarly, the chapters of this volume also express a lively exchange of perspectives among the investigators currently pursuing similar questions with their respective species.

Support for the preparation of this volume and the Primate Cognition Project was provided by the National Science Foundation BNS-9022355 (STB), and laboratory support for EJC from NSF grant BNS-8515831. We also wish to acknowledge the contribution of Michelle Hannan toward the preparation of the indices.

—Sarah T. Boysen
—E. John Capaldi

PART ONE

EMPIRICAL APPROACHES TO COUNTING (OR NUMERICAL COMPETENCE) IN ANIMALS

CHAPTER ONE

Invisible Counting Animals: A History of Contributions from Comparative Psychology, Ethology, and Learning Theory

Mark Rilling
Michigan State University

The title of this chapter has a double meaning. Invisible counting refers both to a cognitive representation of number and to the virtual invisibility of counting in animals as a topic in textbooks written for undergraduates. The chapter is an attempt to explain why the topic has been virtually invisible. Although establishing a date for the beginning of the experimental analysis of animal behavior is arbitrary, 1898 is a good choice. In that year, Thorndike published his dissertation, *Animal Intelligence*, and Romanes reported a successful demonstration of teaching a chimpanzee to count to five. The experimental study of animal behavior is a field in which short-lived fads are the norm. Such problems have a half-life that is really too short for historical analysis.

The question of counting in animals is an exception. For over 90 years, experimental research on counting has been reported from laboratories throughout the world. The subject matter of psychology has undergone three paradigm shifts during this time span: from consciousness, to behavior, and currently rests on cognition. Some of the best experimenters in animal behavior contributed to the development of valid procedures for studying counting. The successful procedures emerged gradually after decades of failure, independently of paradigm shifts. This chapter seeks to identify the historical causes that produced the successful experiments.

TESTING PHILOSOPHICAL MODELS WITH HISTORICAL DATA

"History of" papers are common in psychology. These papers usually trace the development of some problem from its origin to the present. This chapter is different. Philosophers of science, for example, Kuhn (1962), Popper (1968), and Richards (1987), developed a variety of models for explaining scientific progress. These models have normally been tested in the history of physics or biology. If these models are general models of science, not merely special cases for a particular discipline, then the models should have generality to psychology.

This chapter adopts a style and a model of science employed by Richards (1987) in his recent book, *Darwin and the Emergence of Evolutionary Theories of Mind and Behavior*. According to Richards's view, the writing of science history is itself a scientific task involving data, hypotheses, and theory. Richards (1987) employed a natural selection model to account for the emergence of evolutionary theories of mind and behavior. Basically, Richards believed that the best ideas are selected in science in much the same way as the fittest species are selected by natural selection. Richards's approach contrasts with that of Kuhn (1962), who assumed a revolutionary model in which change is cataclysmic. Kuhn purported that reality looks different after the revolution, much as the form of an illusion depends on the set of the viewer. For the reader, a model has the advantage that the author's bias is explicit. For the writer, a model provides a framework for selecting historical facts and proposing the potential causes of scientific progress.

Science is a mixture of competition and cooperation among communities of scientists. A model of scientific explanation requires a unit for historical analysis of these communities. The unit of the individual scientist is too small, whereas the unit of the discipline is too large. Selecting a unit is a problem for the sociology of science. In her book *Invisible Colleges* (1972), a sociological study of the diffusion of knowledge in scientific communities, Crane arrived at the "invisible college" as a unifying theoretical concept. An invisible college is a group of scientists who employ the same paradigm and work on similar problems. A paradigm refers to the theoretical assumptions held in common by a research community. In this chapter the three major invisible colleges that emerge from historical analysis are comparative psychology, ethology, and learning theory. None of these colleges was homogeneous; each contained individual scientists whose basic assumptions and paradigms were quite different. This chapter highlights these differences.

Some philosophers, for example Kuhn (1962), appear to believe that paradigm diversity is a source of weakness. The history of counting in animals supports the opposite assumption. Paradigm diversity was correlated with scientific progress on counting in animals. The task of the individual scientist is to publish an experiment in a journal that the editor judges by a process of peer review to be an original contribution. Originality is more likely when a group of scientists working on a similar problem hold different assumptions and work from slightly different paradigms. Otherwise, all the scientists working on the same problem would duplicate each other by conducting virtually identical experiments.

COUNTING EXPERIMENTS:
THE CUMULATIVE RECORD

For students of business, a familiar graph plots product sales against years. For a typical product, sales begin slowly, rise rapidly, slow down, and eventually fall off. In science, the product is original experiments that are reported as publications in scientific journals. Much like soft drink companies, this chapter assumes that invisible scientific colleges compete for research grants and space in journals. Crane (1972) employed a cumulative record of yearly publications similar to a sales graph to test her theory that the history of a scientific idea progresses through an orderly and predictable series of stages. The cumulative record provides quantitative data for historical analysis. If the life cycle of a scientific idea is similar to the life cycle of a product, then the shape of the cumulative record of publications will be a growth curve. The four stages described here are slightly modified from those of Crane.

In Stage 1, a new research problem appears. In this case, the problem is counting in animals. For outsiders, the value of the problem is not apparent and it is ridiculed by competitors who fail to grasp the significance of the work. Some competitors may even have a hidden agenda, marketing a different product. If the idea were a stock in a corporation, its value in the stock market would be described as undervalued. For insiders, publications during this period become the "classical" papers that are most frequently cited by subsequent investigators. Original theories emerge. This is the stage of peak scientific creativity for which Nobel prizes are properly awarded. Unfortunately, obtaining funds for research at this stage is difficult because granting agencies are likely to be dominated by competitors. Few publications appear, so the slope of the cumulative record is relatively flat.

Stage 2 is a period of accelerating growth produced by a substantial increase in the rate of publication. Valid procedures for measuring the phenomenon appear. Solid, original, but not seminal, work occurs during this stage. Critical control procedures, missed by the pioneers, become a routine part of the experimental designs. The problem is now a hot topic. Symposia are held at scientific meetings. Editors, once cool to the problem, now eagerly solicit papers. The camp followers climb on board. The bandwagon rolls with an outpouring of well-received papers and adequate funding from granting agencies. The major problems emerging from the idea are solved during this stage.

During Stage 3, the cumulative record decelerates because the originality of the idea starts to fade. The product, that is, scientific idea, is now mature; research in the area is no longer considered on the cutting edge of the field. Extensions of the original work are now obvious, so less creative investigators become attracted to the problem. The base of solid knowledge accumulates, but more slowly than before. The problem appears to be progressing to outsiders who have the illusion that the growth will go on forever. The idea is now overvalued like a high-flying speculative stock. New experiments may produce anomalies the original theories cannot explain. The investigators who made the original discoveries have long since moved on to other questions.

Stage 4 is stagnation, the end of the product cycle of originality. The slope of the cumulative record shows substantial deceleration, and ultimately becomes flat as fewer new papers appear. The lode of originality stimulated by the original problem is now exhausted. Creative investigators have long since moved on to other problems. All but the most trivial questions have now been answered. The idea is old hat. Editors reject papers on the topic as uninspired replications of well-established work. Reviewers of grant proposals are unenthusiastic about the prospects for funding work in the area. Camp followers start looking for another bandwagon. The product cycle of the scientific idea is now complete. Unless the members of the invisible college generate new problems, the idea is dead. Historians begin to find the topic interesting because the tools of historical hindsight can now be applied to the problem.

This chapter extends the tool of the cumulative record of publications developed by Crane to the history of counting in animals. Figure 1.1 shows the cumulative number of experimental papers on counting published during the 20th century. Included are all original, experimental papers on the topic found by the author. The decision rule was conservative; the chapter must include some discussion of counting in animals, or a subsequent reviewer must reinterpret the data as an experiment on counting. Excluded is work on the double alternation problem and transitive inference. Included is research on discrimination between different schedules of reinforcement because such work is now cited in contemporary reviews of counting (Davis & Pérusse, 1988; Gallistel, 1988).

All of the papers appeared in journals with peer review. The theory developed here predicts that Stage 1 papers are likely to be rejected by well-established journals. If authors fear that a paper will be greeted with

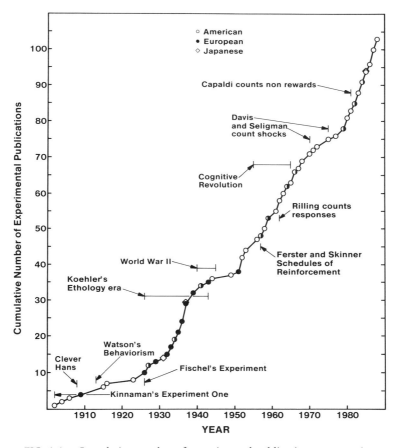

FIG. 1.1. Cumulative number of experimental publications on counting in animals from 1900 to 1988.

ridicule, then they might send the paper to a newly established journal in anticipation of a more hospitable reception. A major reason for establishing a new journal is to create a new outlet for papers with a new orientation. In addition, the first volumes of a journal are likely to contain more significant papers than subsequent volumes. To jump ahead of the story, the first paper on counting in animals was published by Kinnaman (1902) in Volume 13 of the *American Journal of Psychology*. This journal was controlled by G. Stanley Hall, who sponsored Kinnaman's work. In Europe, the *Zeitschrift für Tierpsychologie* [Journal for Animal Psychology] was founded in 1937 as a new outlet for research on ethology. Otto Koehler, the ethologist who is the father of counting in animals, was an editor of that journal from its inception. Many of the most important papers from Koehler and his students appeared in the early volumes of the *Zeitschrift für Tierpsychologie*. The *Journal of the Experimental Analysis of Behavior* was founded in 1958 by Skinnerian behaviorists who were having trouble placing papers in the official journals of the American Psychological Association. Ferster (1958), one of Skinner's most creative students, published a paper of relevance to counting in Volume 1 of the *Journal of the Experimental Analysis of Behavior*. These three journals were founded by individuals who did not agree with each other about the definition and goals of research in animal behavior. Yet in each case, significant papers on counting in animals were published in the early volumes of these journals. At least for the problem of counting in animals, the data confirm the hypothesis that creative papers appear in early volumes of new journals.

Only 103 papers on the topic have appeared between 1902 and 1988, so the average number of papers per year is 1.20. Counting in animals has not been a popular topic among research scientists! Counting is considered here as a cognitive process (i.e., invisible counting). Therefore, one might expect counting to be out of fashion during eras of behaviorism. This is not the case. The slope of the cumulative record showed sharp acceleration during the era between 1925 and 1940 at precisely the time when behaviorism was becoming a dominant view in U.S. psychology (Samuelson, 1985). The literature on counting between 1926 and World War II was dominated by European ethologists. Of the 24 papers published between 1926 and 1940, 19 were published by European ethologists. World War II was a watershed during which basic research virtually stopped.

For most historians of psychology, the most important postwar event was the cognitive revolution, which took place from 1955 to 1965. Investigators were urged to turn toward human cognition as a valid subject matter of psychology. The cognitive revolution was, at least in part, an attack on behaviorism, an outlook well represented in animal learning. Yet this time period showed 19 papers on animal counting, a relatively productive period of research. Many of these papers emerged from laboratories dedicated to operant conditioning in the tradition of Skinner. Initially the cognitive revolution had no impact on the problem of counting in animals. Cognitive concepts like counting were presumably taboo among the Skinnerians, so why did counting experiments emerge from the laboratories of Skinnerian behaviorists? Although the overall slope shows an accelerated rate of publication over the years, the detailed record shows that activity fluctuated over the decades. What are the historical causes of these fluctuations? This is the question addressed in the remainder of this chapter.

"ARE ANIMALS ENDOWED WITH THE FACULTY OF NUMBERS?"

The title of this brief section is a heading from Gall's (1835) six-volume work, *The Influence of the Brain on the Form of the Head*. Phrenology was one of the first systematic programs for determining how the brain controls behavior (O'Donnell, 1985). Gall proposed a list of faculties, each of which was assumed to have an anatomical basis in the brain. He decided not to answer the question whether counting was a faculty and considered an alternative to counting. He proposed that it was not necessary for a bitch to count her pups because she could recognize each individually. Gall's section on counting was followed by a section devoted to the faculty of time. In the case of timing, Gall speculated there was an organ in the brain for timing.

According to one of the themes of this chapter, scientific progress on the problem of counting in animals occurred in a series of small steps. A prerequisite for a good experiment is a good question. In 1810, Gall demonstrated that a scientist could pose the question of counting in animals. However, over 90 years were to pass before the question arrived in the laboratory.

ANIMAL COUNTING IN THE AGE OF DARWIN

The roots of contemporary research on counting in animals spring from folk psychology. Perhaps an ancient shepherd first speculated that his dog counted sheep. Counting in animals was not discovered by scientists. They merely brought the behavior into the laboratory for experimental analysis. The question of counting in animals was on the agenda of the first comparative psychologists. One of the earliest calls for experiments on counting in animals came from Lindsay in 1880 in his book, *Mind in the Lower Animals*. This book included chapters called "Faculty of Numeration," or counting, and "Power of Calculation," or timing. Lindsay's chapter was mainly filled with colorful anecdotes and he concluded that "it cannot be said that the nature and extent of the knowledge of numbers possessed by various animals are yet thoroughly understood, or have been satisfactorily demonstrated" (p. 453). It is fascinating to note that over 100 years later, Gallistel (1990) wrote a chapter with headings similar to those of Lindsay. In Gallistel's chapter, a section on timing is followed by a section on counting. For those who believe that scientific progress is a slow, painful accumulation of knowledge across generations, a comparison of Lindsay's and Gallistel's chapters reveals that the 100-year effort has paid off with an accumulation of knowledge that fully validated Lindsay's call for experimental research.

The two most prominent early comparative psychologists were George Romanes and C. Lloyd Morgan. Darwin was Romanes's mentor and Romanes was Morgan's mentor (see Richards, 1987, pp. 331–408 for a discussion of the complex relationships among these titans of evolutionary thought). Their disagreement about counting in animals was a precursor of the clash between cognitivism and behaviorism. Today Romanes would be called a cognitive psychologist. In *Animal Intelligence* (1883), Romanes proposed a criterion for mind that has a contemporary ring. Mind was an *inference* from the activities of the organism. Which activities counted as mind? Romanes's answer was learning. "Does the organism learn to make new adjustments or modify old ones in accordance with the results of its own experience?" (Romanes, 1883, p. 4). Applying the criterion of learning to counting, Romanes looked for evidence of counting in animals.

Counting arrived in the laboratory in a series of stages. First, anecdotes were collected by naturalists. Next came demonstrations on pets and animals in zoos. Finally, experiments were carried out by scientists who controlled the life history and experimental conditions in their own laboratories.

The transition in methodology is clear in Romanes's (1898) third book, *Mental Evolution in Man: Origin of Human Faculty*. Anecdotes about birds who counted the number of hunters who entered an ambush appeared in books on natural history during the 18th century. Gall (1835) quoted a version of a story about a magpie in which it was necessary for five or six men to enter an ambush in order to trick the bird into forgetting that one hunter remained behind to shoot the bird after the others left. Romanes cited a similar version of a counting crow from a book written by the ranger at Versailles whose job included shooting this pest. The ranger discovered that the crow would not return to its nest during daylight if it saw a hunter entering a blind. The independent variable in this study was the number of rangers entering the blind. The strategy was for a party of men to enter the blind and subsequently leave the hunter behind. Romanes (1898) reported that "it was found necessary to send five or six men to the watch-house in order to put her out of her calculation" (p. 57). Romanes did not consider this anecdote evidence of counting. Today, his interpretation would be called a relative numerosity judgment (Davis & Pérusse, 1988), in which the bird perceived that more men went into the hut than left it.

Romanes (1898, p. 58) also reported that he taught a chimpanzee to count correctly to five. On the oral request of a number by Romanes, the animal handed him straws held in her mouth. Romanes had no doubt that the animal discriminated among the numbers 1–5 and understood the name for each. However, as Kinnaman (1902) pointed out, such cases could be explained as an association between the verbal cue and the number of straws without appealing to the idea of number.

In 1895, Morgan wrote a book called *An Introduction to Comparative Psychology*. The title was ironic because Morgan did not believe that a truly comparative psychology was possible (see Richards, 1987, pp. 379–381). He was not willing to infer consciousness from behavior. Instead Morgan advocated "objective psychology," a study of the adjustive behavior of animals (Richards, 1987), that is, a precursor of behaviorism. Morgan started from the assumption of a fundamental difference between human and animal consciousness, thereby rejecting the Darwinian assumption of continuity. Romanes was Morgan's mentor. In return, Morgan became Romanes' most insightful critic. In Morgan's view, animal behavior was not intentionally directed toward a goal. Animals were stuck with instinctive behavior, sensory experience, and associative learning, and language was required for cognitive processes like counting.

Morgan was in the tradition of skepticism toward counting in animals and this tradition continues to the present day. A skeptic explains counting in terms of alternative processes, processes considered simpler. Morgan (1895) did it well, as the following quote indicates: "One can very readily distinguish a succession of three from a succession of four, without anything like counting, through the sensing of sense-experience" (p. 232). Animals did not perceive the number of objects by counting. If a number of objects are lined up in a row in front of a subject, the problem can be solved by running the eye along the row, and responding on the basis of the duration of the scan. This is an example of Morgan's (1895) famous canon: "That in no case is an animal activity to be interpreted as the outcome of the exercise of a higher psychical faculty, if it can be fairly interpreted as the outcome of the exercise of one which stands lower in the psychological scale" (p. 53). In the case of counting, Morgan reduced the problem of counting to a problem of timing. There is a problem with Morgan's example, however, because there is no reason to suppose that timing is a simpler process than counting.

Morgan's canon has never worked well in animal behavior. What criteria does a theorist employ to decide if an explanation is simple or complex? Morgan provided no guidance. One can argue that all psychological processes are complex in the sense that many independent variables control all the psychological processes investigated to date. Moreover, with Morgan's canon the relative complexity of two processes may change in the course of scientific history. Timing was simple, but counting was complex. Gallistel (1989, 1990) placed counting and timing at the same level of complexity; each is considered a relatively simple process. Counting lost its magical complexity when Gallistel (1988) wrote, "Counting is an extremely simple process" (p. 586).

Morgan established the role of scientific critic as a fixture in experimental psychology. The history of counting in animals is a story of conflict between scientists and their critics. The critic's job was to say that an experiment was not perfect and to suggest alternative explanations. Counting is easily confounded with other sources of stimulus control. A critic could simply select a plausible stimulus from the environment that could explain any experiment on counting in animals. Failing plausibility, the desperate critic could reach for an implausible alternative to counting. It was not always necessary for the critic to run an experiment actually demonstrating control by the alleged stimulus. Other critics of the counting hypothesis included Pfungst in Europe and Watson in the United States. Across decades, the goad of criticism produced a rising standard of scientific excellence.

At this point the reader, quite understandably, might like to have a definition of counting. However, operational definitions did not arrive in the social sciences until the 1930s. The following definition is, at least, implicit in the early papers and consistent with contemporary usage (Gallistel, 1988). In *counting*, a discrimination is learned on the basis of the number of events, rather than on the basis of some other stimulus correlated with the number of events. Excluding these alternative sources of stimulus was a task requiring decades of painstaking research.

THE SIDESHOW OF "CLEVER HANS"

Clever Hans, the "clever" horse that could not count but whose tapping responses were visible to the world, is perhaps the most famous subject in experimental psychology. His owner was Wilhelm von Osten, a high school mathematics teacher who trained Clever Hans to count. Von Osten convinced the press and some Darwinian evolutionists in Germany that Clever Hans was actually counting. The leading psychologists in Germany regarded all this hoopla as hocus-pocus. Pfungst was appointed to investigate von Osten's claims. He concluded that Hans was responding to small head movements from his questioner. Pfungst's book, *Clever Hans*, was published in an English translation from the original German in 1911, accompanied by a warm preface written by James Angell, Watson's mentor (see Rosenthal, 1911/1965). Hans was not the first animal trained to appear intelligent. Mountjoy and Lewandowski (1984) provided examples from the 17th and 19th centuries.

The translation was reissued (Rosenthal, 1965) with an excellent introduction by Rosenthal, who warned investigators about the problem of biased data produced by unintended communication between the experimenter and subject. In 1981, the Clever Hans story, now a "phenomenon," was a topic for a conference and subsequent book sponsored by the New York Academy of Sciences (Seboek & Rosenthal, 1981). This book is concerned with the "Clever Hans Phenomenon," that is, unintentional communication between people and other animals. Clearly animals can be trained by operant conditioning techniques to appear intelligent. However, such pseudo-counting is irrelevant to the question of the current chapter. Here we want to determine whether counting, the cognitive process, is a component of animal intelligence. None of the chapters in Seboek and Rosenthal's book deals with the question of counting in animals. "Clever Hans" now stands for experimenter bias and as such, has become an icon for experimental psychologists. From the standpoint of laboratory research on counting in animals, Clever Hans was a sideshow. He straddled the fence between anecdote and the laboratory. He was trained by an amateur but was tested by a professional. From the perspective of this chapter, the most insightful comment about Pfungst's research with Clever Hans was written by Hediger (1981), an animal psychologist. According to Hediger's view, Pfungst was just as guilty of wishful thinking as von Osten, because Pfungst's obsession was the belief that animals do not think like humans. Curiously, rather than conducting experiments with laboratory animals, most of the experimental data in Pfungst's book came from experiments on the detection of small head movements by humans. For research on counting, the legacy of Clever Hans was the realization of the necessity of eliminating noncounting interpretations of the experimental data. To his credit, Pfungst closed his book with a call for additional research on the problem of counting in animals. Figure 1.1 shows that, although several laboratory experiments preceded the publication of Clever Hans, most laboratory experiments followed it. Thus, Clever Hans was not entirely irrelevant to the laboratory history of counting in animals. His legacy ensured that investigators added controls for noncognitive sources of stimulus control and became much more conservative in interpreting data from counting experiments.

COUNTING ENTERS THE LABORATORY AS COMPARATIVE PSYCHOLOGY

Comparative psychology crossed the Atlantic just in time to help G. Stanley Hall, the president of the newly established Clark University, with a major political problem. Hall was an academic entrepreneur par excellence who founded the American Psychological Association and *The American Journal of Psychology*, and who jumped from chairing the psychology department to heading the university. Hall promised a tradition of research in the psychology department at Clark, but by 1900, the department was filled with a "lunatic fringe" of graduate students whom Clark himself called "crazies" (O'Donnell, 1985, p. 163). Clark was also an enthusiastic Darwinian. He introduced comparative psychology as a rigorous, new, laboratory program in the psychology department. This program quickly caught the attention of students who saw Darwinian research with laboratory animals as an opportunity to launch a career as a professional psychologist from a platform of scientific respectability. Hall suggested the project that became the first systematic, experimental investigation of counting in animals. The lucky graduate student who got the counting project was A. J. Kinnaman, a Fellow at Clark University. Interestingly, none of the doctoral graduates from Clark's comparative psychology program used the degree as a springboard to an academic career in psychological research. After running the first systematic laboratory experiment on counting in animals, Kinnaman became the head of a school for teachers (O'Donnell, 1985, p. 165).

The history of laboratory experiments on counting in animals begins with these words, "There has been considerable written but very little done toward a rigorous examination of the number notions of lower animals" (Kinnaman, 1902, p. 173). Kinnaman (1902) and Porter (1904) each addressed the question of counting in animals. Kinnaman chose the rhesus monkey whereas Porter selected the English sparrow. In the counting experiments, the experimenter presented the subject with a series of bottles arranged horizontally in a row. Food was placed in one of the bottles and the percentage of correct responses was observed at various locations. Kinnaman interpreted his results by following Morgan. He concluded that his subject did no real counting but had learned to respond correctly on the basis of the position of the correct object. Porter also obtained negative results for the counting hypothesis, also preferring an interpretation in terms of a position habit. Thus the experimental tradition of research on counting in animals started with a conservative interpretation of the data. The problem moved from the realm of anecdote into the laboratory with two experiments from which essentially negative results were obtained. It should come as no surprise that the first two reports of laboratory investigations of counting in animals were published in Hall's journal, *The American Journal of Psychology*.

In the United States, Watson and Yerkes replaced Morgan and Romanes as protagonists. Unlike Hall, Watson was hostile to Darwinism. In a letter to Yerkes by Watson dated October 29, 1909 he wrote "Damn Darwin" (O'Donnell, 1985, p. 159). When Watson arrived at Hopkins from Chicago, the university had given the newly established department of psychology the mission of serving the local community, specifically offering courses for local teachers. Watson considered himself a comparative psychologist, but his brand of comparative psychology was free of evolution. In learning theory Watson found a product that justified research on laboratory animals that could be sold to the students he was required to teach. Learning theory was a topic that could be offered in psychology courses designed to appeal to teachers. No other academic department competed with psychology for offerings in learning theory. Of course, behaviorism was an intellectually honest enterprise, but it was also a product of the academic pressure Watson encountered at Hopkins.

During the winter of 1913, Watson gave a series of eight lectures on behaviorism at Columbia University. These lectures were expanded into a book called *Behavior, an Introduction to Comparative Psychology*. In many ways the book represented a regression to pre-Darwinian thought about behavior. For example, man was considered unique because of his language capacity. In a return to the view of Descartes, Watson wrote that the lack of language habits differentiated people from brutes. Animals were referred to as "beasts." Intelligence and most of the topics previously covered in books on comparative psychology were missing.

Comparative psychologists never agreed among themselves about the proper topics for the discipline. Romanes's primary concern was animal intelligence. Morgan focused on instinct. Hall leaned toward development. For Watson, animal learning was the key to unlocking the mysteries of animal behavior. Watson proposed animal learning as the single most important topic in the whole study of behavior, and his book was filled with learning curves and diagrams of experimental apparatus. However, missing from Watson's book was a discussion of the serious laboratory work that had been carried out on comparative cognition. One exception was a discussion of Hunter's (1913) work on the delayed reaction. Hunter observed that raccoons responded correctly across a delay of 25 seconds even though the body orientation of the animal could change during the delay. Hunter was able to explain the behavior of rats and dogs as mediating behavior because those animals could solve the task only by orienting toward the correct goal box during the delay. Watson was grasping at straws when he speculated that the raccoons in Hunter's experiment might be responding to olfactory cues given off by a light bulb! Within Watson's theory a response was always preceded by a stimulus in the environment, so the task of the experimenter was to find that stimulus. The idea that the animal maintained a representation of the stimulus or rehearsed it was beyond the pale. To his credit, Watson reserved judgment about Hunter's raccoons.

Most interesting was Watson's discussion of Kinnaman (1902) and Porter (1904). Watson presented learning curves from these studies and pictures of the apparatus, but he made no mention of the research on counting. Instead, his chapter on the "Limits of Training in Animals" was filled with a discussion of Clever Hans. Even though he could point to the negative results of Kinnaman and Porter, Watson preferred the cautionary tale of Clever Hans. As a matter of experimental strategy Watson advocated studying learning first, then moving on to more complex types of behavior. His rationale was to lay a firm foundation for the science of behavior. Just to make sure his reader got the point, Watson (1914) wrote, "From our point of view, it can readily be understood the search for reasoning, imagery, etc. in animals must forever remain futile, since such processes are dependent upon language" (p. 334). Watson never became a major learning theorist. After World War II, the U.S. learning theorists who developed procedures for investigating counting followed Watson's strategy of studying simple learning first, and procedures for studying counting emerged slowly from the foundation of learning theory.

Some writers suggest that Watson's views revolutionized psychology immediately. Watson knew better. As the storm of his impending divorce led him to realize that he would be forced to leave academia he wrote, "I feel that my work is important for psychology and the tiny flame which I have tried to keep burning will be snuffed out if I go-" (letter from Watson cited by Boakes, 1984, p. 224). One of many psychologists who did not find Watson's flame illuminating was Yerkes.

Yerkes was a Darwinian in the tradition of Romanes. He tipped his hand in his *Introduction to Psychology* when he wrote, "The more liberal among psychologists are at present inclined to believe that at least some animals . . . experience some conscious complexes which are much like ours" (Yerkes, 1911, p. 239). One "complex" that interested Yerkes most was counting. Following the lead of Kinnaman and Porter, Yerkes (1916) designed a multiple-choice apparatus for studying counting in a variety of different species.

The heart of the apparatus was a series of nine compartments. The animal had to pass through the correct compartment to obtain food. Each compartment had an entrance and exit door that could be opened by a rope pulled by the experimenter. In previous work the physical location of the correct compartment was held constant. Yerkes varied the physical location of the correct compartment from trial to trial, but reinforcement was based on the relative location. This was accomplished by opening a different set of doors at the beginning of each trial. Suppose reinforcement occurred for selecting the second entrance door from the right. If the entrances to boxes 5–9 were open, then the correct choice was 8. If boxes 1–3 were open, the correct choice was box 2. However, this task was simply too difficult. After years of research, first with pigs and then with primates, Yerkes gave up discouraged and disappointed (Coburn & Yerkes, 1915; Yerkes, 1916; Yerkes, 1934; Yerkes & Coburn,

1915). World War I disrupted research in comparative psychology. After organizing IQ testing for the army during World War I, Yerkes abandoned research to become a successful administrator. Spence (1937) inherited the use of the Yale primate colony, but even his results in the counting domain were meager.

In the United States those investigators who followed Watson's advice to study learning in laboratory animals found themselves riding on top of the field. The era prior to World War II was the golden age of learning theory in which the leading theorists (e.g., Guthrie, Hull, Spence, Skinner, and Tolman) found themselves at the pinnacle of psychology. Those who followed Yerkes's effort to establish counting in animals found themselves on the eve of World War II looking at negative results, notwithstanding 40 years of effort.

OTTO KOEHLER: COUNTING AN INVISIBLE ETHOLOGIST

Where the U.S. scientists failed, a German ethologist succeeded. This was Otto Koehler (1889–1974). He conducted and supervised a series of programmatic experiments that finally established counting in animals, that is, the representation of number as a cognitive process. In 1936 with Konrad Lorenz, he was the founder and coeditor of the *Zeitschrift für Tierpsychologie* [Journal for Animal Psychology]. Today his name remains on the masthead of that journal, now called *Ethology*. His vita (Hassenstein, 1974) lists 189 publications from 1912 to 1974. In addition, he supervised 64 publications by students. The scope of his research program covered a variety of species and included topics such as homing and sensory processes. However, the jewel in the crown was his studies of counting in animals.

Beginning in 1926 with a student paper by Fischel (1926), culminating in a brilliant theoretical review (Koehler, 1943), and ending in 1957 with a nostalgic paper in English (Koehler, 1957), the goal of Koehler's research program was an ethological analysis of the counting process. This research program spanned 30 years and produced 17 primary research papers on counting in animals—far more than that of any other investigator. In 1973 the Nobel prize was awarded for research in ethology to Lorenz, Tinbergen, and von Frish for their contributions to the genetic determinants of behavior. Only three individuals can share a Nobel prize, so Koehler was overshadowed by three more luminous figures. Koehler was also a founder of ethology, but his interest in learning was atypical for an ethologist. Atypical behavior from a member of a scientific community crops up often in this history. Today, in textbooks of comparative psychology or ethology, the topic of animal counting is rarely mentioned nor is the name Otto Koehler. But how could a man whose name was so prominent in the *Zeitschrift für Tierpsychologie* remain invisible to so many scientists whose interests were animal behavior? Why was Koehler overlooked by the textbook writers?

On the occasion of the 60th birthday of Konrad Lorenz in 1963, *Ethology* published a set of papers on the aims and methods of ethology. In one of these papers, Tinbergen (1963) wrote, "Ethologists themselves differ widely in their opinions of what their science is about. . . . It is just a fact that we are still very far from being a unified science, from having a clear conception of the aims of study, of the methods employed and of the relevance of the methods to the aims" (p. 410). So much for those philosophers who assume that an area of research needs a single paradigm! Ethologists did not agree about the definition of their subject matter beyond a loose definition of ethology as the biological study of behavior.

Koehler (1957) followed Darwin and other ethologists by assuming continuity between human and animal behavior. However, two interests set Koehler apart from other ethologists. They were primarily interested in instinctive behaviors that could be readily observed in the field. In contrast, Koehler was interested in learning and thinking without language. He worked in the laboratory on a problem that many of his fellow ethologists must have considered unnatural. Koehler was an unusual ethologist who studied an artificially learned behavior in the laboratory that he believed did not occur in the wild. By selecting counting as a problem that required a long training period of trial-and-error learning before the test for counting could occur, Koehler's research strategy was more similar to that of U.S. learning theorists than to that of his European colleagues in ethology. He was repelled by popular reports of counting in clever animals, reports he regarded as quackery and humbug. Recall that the question of Clever Hans was posed by Pfungst as a question of animal consciousness. Koehler (1957) believed that he had nothing to say about consciousness. For Koehler, counting was a process learned without words. His use of the word counting was a description of the animal's carefully observed behavior. Having removed himself from philosophical issues, Koehler went on to address the scientific questions. A prerequisite for studying a learning process is a procedure for validly measuring the process. Prior to 1926, alternative explanations other than counting were always plausible. Before Koehler's research program, the U.S. comparative psychologists had spent about 25 years working on the task, yet they failed to develop a valid preparation for studying the process of animal counting in the laboratory.

The intellectual antecedents of Koehler's research were Darwin's continuity hypothesis and Wolfgang Köhler's research methodology. Köhler's research on animal intelligence was carried out within the framework of Gestalt psychology. While U.S. psychologists were turning away from intelligent behavior under the banner of behaviorism, Köhler was busy demonstrating how to study cognitive processes in animals. Köhler was far more successful than Yerkes in developing procedures for studying animal intelligence. His classic volume, *The Mentality of Apes* (1925) provided a roadmap for investigators who wanted to study animal cognition. Koehler followed Köhler's strategy of presenting the animal with a problem and observing the animal until the problem was solved. Koehler's goal was to study behavior as a process involving the whole organism at a time when the dominant biology of the day consisted of examining tissue under the microscope or embalmed specimens under glass.

In their paper on numerical competence in animals, Davis and Pérusse (1988) classified the procedures for studying counting in animals into two categories depending on whether the events to be counted are encountered simultaneously or successively. These are also the two capacities that emerged from Koehler's research. He referred to the simultaneous procedure as "seeing numbers" and called the second capacity, in which the animal had to repeat an act a certain number of times, "acting upon a number."

By 1915, Köhler developed a procedure for investigating sensory processes of vision in the chimpanzee and pigeon. An experiment in 1926 by Fischel borrowed Köhler's methodology and established the simultaneous counting procedure still in use today. The stimuli were grains of corn or small rocks glued to pieces of cardboard. On each trial the pigeon was presented with a choice between a positive and a negative sign, each of which represented a number. The pigeon was placed on a table on which the discriminative stimuli were located. After the pigeon went to one of the objects, the experimenter raised a door behind which food was located if the pigeon had chosen correctly. The pigeons were first trained to discriminate 2 (+) from 1 (−) or 3 (−). After mastering this task, the birds were trained to discriminate 3 (+) from 2 (−) or 1 (−). The birds readily mastered these simple tasks.

Fischel succeeded where the U.S. scientists failed. His experiment laid the foundation for Koehler's subsequent work. However, the performance of Fischel's birds was fragile, and the discrimination disintegrated when the slightest change was made in the "Gestalt" or configuration of the objects forming the number. Fischel's animals were probably discriminating form, not number, but this contemporary interpretation is beside the point. Koehler's laboratory became the center for research on counting in animals, and subsequent work added elegance and appropriate controls.

The basic problem with the simultaneous procedure is separating stimulus control by a feature of the stimulus from stimulus control by number. The problem is solved by holding the number of objects constant throughout the experiment while varying the size of the objects from trial to trial. This control by number was achieved by Koehler (1943) with a raven named Jacob, who was first trained with the simultaneous procedure. Jacob's task was to choose the pot with 5 points on its lid from among a set of 5 pots, each of which had a different number of points on each lid. During training the size and form of the spots for the concept of number was randomly varied from trial to trial. In the final test for transfer, the number of objects remained constant, but the size of the objects was varied across a range from 1 to 50 units of surface area. For the transfer test the objects were flat chunks of plastic broken into highly irregular pieces. Jacob solved the problem by selecting the positive stimulus thereby displaying concept learning by the particular number.

The successive procedure is closer to the intuitive layman's concept of counting as repeating a sequence of behaviors and stopping at the correct number. Here the task of the organism is to retain a changing representation of the successive count in short-term memory, and the task of the experimenter is to make sure that the resulting behavior is not confounded with time or any other extraneous stimulus.

Arndt (1939) illustrated the successive counting procedure with pigeons. The bird's task was to eat only 5 peas from a cup and then to walk away from the cup, at which time another trial began. The peas were delivered from a pellet dispenser and the experimenter manipulated the intertrial interval within a range from 1–60 s. Only one pea was visible to the bird. The number of pecks is not the discriminative stimulus because after landing in the bowl the pea spun around and sometimes the bird made as many as 14 pecks to retrieve that single pea! During acquisition, a punishment procedure was used following errors. The bird was shooed away from the cup if it lingered until the sixth pea was delivered. Later, shooing was automated with a spring-loaded device that looked like a broom. During "spontaneous" tests for transfer, the punishment contingency was removed and the behavior of the organism was filmed without intervention from the experimenter. Schiemann (1939) extended the experiments to jackdaws.

Koehler (1957) reported that the birds performed faultlessly for sequences of up to nine consecutive trials where a trial consisted of eating the correct number of pellets, walking away from the cup, and then returning for the next trial. The probability that such sequences would occur by chance was, of course, very low. There was a constraint on the successive counting procedure, however, because reducing the number, for example from 5 to 4, produced little positive transfer so that each new number required about as much training as the first task.

Koehler's strategy was to employ a wide variety of different methods for each of the two basic procedures. These included matching-to-sample in which the correct number, the sample stimulus, varied from trial to trial. The sample stimulus was a lid lying on the ground without a pot beneath it and the test stimuli were the usual pots with lids. Jacob (Koehler, 1943) performed proficiently when the number of points on the lid ranged from 3 through 7 even though the sample changed from trial to trial. In oddity learning tasks, 4 of the 5 pots were identical and the task of the bird was to select the odd pot.

Among the successive methods was presenting the bird with two piles of grain. The task of the bird was to select a total of 5 grains from the two piles. Koehler (1941) showed a "before" photograph with 3 grains in one pile closest to the bird and 20 grains in a nearby pile. During the trial the bird ate 3 grains from the smallest pile and 2 grains from the larger pile for a total of 5 grains. This counting process is somewhat similar to adding 3 and 2. In another successive task mealworms were located in rows of jars. Some jars were empty whereas others contained 1, 2, or 3 mealworms. By employing a punishment procedure in which the animal was shooed away from the jars if it tried to remove too many mealworms, the animal learned to remove 5 baits from the jars and then walk away when the number was reached. These data led Koehler (1951) to conclude that an "inner token" was responsible for the bird's behavior. Koehler assumed that the representation of each number was identical so that the animals were not really counting with a qualitatively different representation for each number, as is the case when humans think "1," "2," "3."

Over the years, eight species were employed in the counting project. The goal was to find the upper limit of counting on the simultaneous and successive tasks. For each species the simultaneous and successive operations converged on the same number. The limit was 5 for pigeons, 6 for budgerigars and jackdaws, 7 for Amazones, magpies, grey parrots, people, ravens, and squirrels. The research with humans was carried out tachistoscopically with a brief duration, and the limit of 7 hit the magic number so familiar to U.S. cognitive psychologists.

Koehler did not believe that his results represented true counting because all his animals showed no transfer from a trained number to a new exemplar; each new number discrimination required relearning. He concluded that, unlike humans, no animal ever gave a name to a number. He viewed wordless counting as a type of concept learning that provides an internal representation of the number of objects in the outer world much like the concepts of greater and smaller, up and down.

The filled circles in Fig. 1.1 clearly demonstrate that the European ethologists were primarily responsible for the accelerated interest in counting in animals that peaked prior to World War II. Of the 27 papers published between 1926 and 1945, 10 were papers of Koehler and his students. Koehler's research program on counting produced nearly 650 pages of published work, most appearing in the *Zeitschrift für Tierpsychologie*. The quality of the research was much more sophisticated than Skinner's initial research on stimulus control in pigeons (see the data in Skinner, 1965), which was collected during World War II and therefore is contemporary with Koehler's research. Koehler's research was acknowledged by all the reviews of animal counting (Davis & Memmott, 1982; Davis & Pérusse, 1988; Honigman, 1942; Salman, 1943; Wesley, 1961), so the work was not unknown. Why then has it been undervalued?

The first problem for English-speaking scientists was the language barrier. Koehler's first paper in English was published in 1950. It was a translation by W. H. Thorpe of an address given by Koehler prior to a film demonstration at the famous 1949 meeting in Cambridge, England, at which the Dutch, British, and German ethologists made peace with each other after World War II. This paper (Koehler, 1951) did not contain data, but subsequent descriptions by reviewers of Koehler's work as "anecdotal" is patently unfair. Most of Koehler's papers (e.g., Koehler, 1943) were published in German in the *Zeitschrift für Tierpsychologie*. Some of the most important papers were published during World War II in volumes likely to be missing from the shelves of all but the most comprehensive research libraries. At first, reviewers (e.g., Salman, 1943) were skeptical of the results, but later reviews (e.g., Davis & Memmott, 1982) acknowledge that Koehler was successful in teaching a variety of species to count. Matching-to-sample and oddity-from-sample have only recently become standard techniques for research in animal cognition, yet Koehler's birds were proficiently performing such tasks prior to World War II. It seems likely that good ideas for research on counting in animals still remain to be discovered in the papers of Koehler and his students. For example, Emmerton's (1988) research on the question of whether pigeons interpolate a numerosity discrimination demonstrates that contemporary investigators may still draw on Koehler's legacy.

Consider the following explanation for the invisibility of Koehler's work in textbooks: The task of the textbook writer is to simplify a field and to sharply distinguish field x from field y. The task of the scientist is to conduct original work that is likely to redefine or fall outside the categories locked into chapters in undergraduate texts. Koehler was an ethologist interested in learning and cognition. Ethology texts are strong on species-specific behavior and weak on learning and cognition. Animal learning texts are strong on learning, but weak on comparative psychology and especially likely to overlook work by non-Americans. Cognition texts are often so egocentric that the reader could complete the course

1. INVISIBLE COUNTING ANIMALS   23

with the impression that only Homo sapiens are sapient! Koehler's research program on counting in animals simply fell between the cracks used to define the traditional fields of psychology and animal behavior.

THE LEARNING THEORISTS LEARN TO COUNT

So far the history of counting in animals appears to be a story of continuous progress. After the U.S. comparative psychologists fumbled, the European ethologists grabbed the ball and made substantial progress toward the goal of establishing valid procedures for studying counting processes. World War II produced a break in scientific progress. Many postwar papers do not cite prewar work. After World War II, the history of counting starts anew almost from scratch. Figure 1.1 shows that the decade between 1940 and 1950 produced only five papers, a return to the rate between 1900–1910 when this history began.

Koehler's laboratory was at Königsberg in Prussia. During World War II, his laboratory was fire-bombed and his archival films destroyed. In 1945, Koehler fled the Russian army's invasion of Prussia with two suitcases and his violin. He was unable to resume his research program until 1949. Even so, U.S. researchers were at most only dimly aware of Koehler's ethological research program on counting.

After World War II, the pendulum of research on counting swung back to the United States. The decade of the fifties is a key period in the history of experimental psychology in the United States. The period between 1955–1965 marks the "cognitive revolution" when psychologists developed procedures for studying cognitive processes in humans. The same period also marked a renaissance in animal learning. Researchers in animal learning finally followed Watson's admonition to investigate problems in learning theory, rather than to investigate cognitive processes. However, the cognitive revolution did not begin to have an impact on animal learning until about 1970. Thus, none of the experiments in Fig. 1.1 that were published during the cognitive revolution were influenced by that event.

The publication of Ferster and Skinner's *Schedules of Reinforcement* (1957) and the founding of *The Journal of the Experimental Analysis of Behavior* (JEAB) in 1958 demonstrated an attempt by Skinnerian behaviorists to dominate the field of animal learning. This journal had been set up by Skinner when the orthodox journals of the American Psychological Association refused to publish papers by his students (see the November 1988 issue of *JEAB* for a set of reminiscences by the scientists who founded *JEAB*). By automating experiments and introducing operant conditioning as a topic for experimental analysis, Ferster and Skinner's research program kept many operant laboratories humming for decades. Of the 103 papers on counting in animals, 7 were published in *JEAB*. The Cumulative Index to *JEAB* for the period between 1958–1973 contained nine papers for which counting was a keyword. The authors of these papers were behaviorists who did not consider counting to be a cognitive process. Instead, they wrote that the number of responses functioned as a discriminative stimulus.

Ferster and Skinner's research program was radical behaviorism, the hallmark of which was *no cognitive concepts*. Yet Ferster's first paper in *JEAB* was a study of counting in the chimpanzee (Ferster, 1958). After leaving Harvard, Ferster went to the Yerkes laboratory in Orange Park, Florida, and one can only speculate about whether he was influenced by Yerkes' work on counting. One of Ferster's major projects was an effort to teach arithmetic to animals (Ferster, 1964; Ferster & Hammer, 1966). The results from this project were disappointing, probably because the authors first attempted to teach the subjects a synthetic language.

Most other learning theorists distanced their research on counting from the animal-language controversy. Of course, the papers in *JEAB* were written in the behavioristic vocabulary of the day, but these radical behaviorists laid a foundation for subsequent research on counting. Figure 1.1 shows that the 1960s marked an acceleration in the cumulative rate of publication of research in counting, for the first time showing real growth in productivity from the peak period of the 1930s. Counting in animals is a cognitive process, so one might have expected that the cognitive revolution would have been the cause of the increase in research on animal counting. It was surprising that the rediscovery of counting occurred in a most unlikely place, the laboratories of the learning theorists. In such laboratories, unobservable cognitive concepts like counting should have been taboo. Unlike Koehler, these investigators did not begin their research with the intention of discovering counting. Counting emerged almost by accident as an interpretation of the data from experiments designed to investigate basic processes in animal learning.

Three examples of the accidental rediscovery of counting in the animal learning laboratory are presented. First, the author's research on counting emerged from research investigating the number of responses as a stimulus on schedules of reinforcement. Next, Davis's research on counting emerged from research on conditioned suppression. Finally, Capaldi's research emerged from research investigating the partial reinforcement extinction effect.

A Personal Sojourn

First consider the evolution of my own research program. I arrived at Maryland in September 1960 with an interest in extrasensory perception. During a summer term away from Oberlin College, I studied as an undergraduate with J. B. Rhine at the parapsychology laboratory at Duke University. Once at Maryland, the interest in parapsychology rapidly disappeared. With Kennedy in the White House, the prospects for research on the experimental analysis of behavior seemed as unlimited as the goals of the space program. I had the good fortune to enter Maryland during the salad days of Skinnerian behaviorism. The *Journal of the Experimental Analysis of Behavior* was publishing Volume 3. My mentor was William Verplanck, who had come to Maryland after an association with Skinner at Harvard, and a large psychopharmacology laboratory was administered by Joseph Brady. Although the focus of the research was not very theoretical, this research group put into practice Watson's idea of predicting and controlling behavior in a laboratory setting. There was lots of grant money and many papers from Maryland filled the pages of *JEAB*.

In January 1962 I left Maryland as a Skinnerian behaviorist with a master's degree and headed for Austin, Texas where I planned to earn a doctoral degree from the University of Texas. In Texas I wound up in the psychoacoustics laboratory where my mentor was Lloyd Jeffress. Jeffress ran a defense research laboratory with a grant from the Navy where the applied work was on undersea warfare.

In the psychoacoustic laboratory we ran experiments on ourselves as subjects. We wore headphones, with the task being to detect a tone embedded in noise. Usually the tone was difficult to detect. The theory of signal detection provided a tool for separating sensitivity to the stimuli from the response bias of the subjects (Swets, 1961). The Texas psychoacoustics group was extending the theory of signal detection to audition, so it was natural to extend the theory to the detection of stimuli resulting from different reinforcement schedules.

At Texas, assistant professors Colin McDiarmid and Charles Watson were my primary mentors. Watson was interested in the theory of signal detectability, which was replacing the idea of the sensory threshold. McDiarmid wanted to set up an operant conditioning laboratory in the Skinnerian model he had seen as an assistant professor at Columbia University. Therefore, it was natural that my first paper in animal learning would combine ideas from the theory of signal detection with basic research on schedules of reinforcement with pigeons.

In a fixed-ratio schedule, reinforcement is contingent on the completion of a fixed number of responses. Ferster and Skinner's (1957) research on mixed schedules included an example of a mixed FR 30 FR 180 schedule in which these two schedules alternated without the presence of an exteroceptive stimulus. The result was a pause on the longer schedule at the point on the ratio where the count was slightly in excess of the smaller ratio. In discussing these data, Ferster and Skinner pointed out that the number of responses could function as a discriminative stimulus. The question was how to measure the accuracy of this discrimination. Our solution (Rilling & McDiarmid, 1965) was to employ two ratio schedules as stimuli and to define the larger ratio, FR 50 as the "noise" and define the smaller ratio as the "signal."

The basic apparatus was a three-key Skinner box. A trial began with the illumination of the center key on which the stimulus ratio was presented. The pigeon pecked at the center key until the stimulus schedule was completed. Then the two side keys were illuminated. Reinforcement of a peck on the side key was contingent on discriminating which schedule had been in effect on the center key. The performance of the pigeons was surprisingly accurate. Pigeons could discriminate FR 35 from FR 50 with an accuracy of greater than 80% correct and at FR 45 the accuracy was greater than 60%. Schedules of reinforcement functioned as stimuli. I considered the possibility that the pigeons were counting, but rejected writing about it. Instead the paper said, "The paper makes no attempt to specify the nature of the discriminative stimuli" (Rilling & McDiarmid, 1965, p. 526).

After the success on ratio schedules, the next step was to extend the theory of signal detectability to the case of the fixed-interval schedule. In a fixed-interval schedule, the first response after a fixed interval of time is reinforced. The task was to determine whether the controlling stimulus was time or the number of responses, two variables normally confounded on a fixed-interval schedule. The number of responses during an interval is not specified by the contingency and varies from one interval to the next.

An FI 45s interval was employed as the noise, whereas shorter values were employed as the signal. The number of responses, not the duration of the interval, turned out to be the controlling variable. The distributions of the number of responses on the short and long intervals were analyzed in terms of the theory of signal detectability. If the animal were discriminating between two intervals on the basis of the number of responses, then as the shorter interval becomes longer, the discrimination should begin to disintegrate at the point where the distributions of responses from the two schedules begin to overlap. Errors should occur when the short interval contains many responses and the long interval contains few responses.

Rilling (1967) computed the mean number of correct and error responses from the distributions for the short and long intervals. Errors were most likely when the birds emitted a larger than average number of responses on the shorter interval and a smaller than average number of responses on the longer interval. The birds behaved exactly as predicted by the theory of signal detectability. It appeared as if the bird were employing a criterion based on the number of responses, which determined whether the bird pecked the left or right keys. When the number exceeded the criterion, the animal pecked the one choice key. When the number was less than the criterion, the animal pecked the other choice key. The birds had been trained to peck the center key only when it was illuminated, thus it was possible to experimentally control the number of responses per interval by darkening the interval for progressively longer durations, which eliminated the confounding between time and number of responses. As the number of responses on the long interval fell, the birds were more likely to make errors as if the long interval was the short interval. Therefore, the number of responses was the controlling stimulus in this discrimination.

I interpreted the data on discrimination between ratio and interval schedules as a number discrimination. The word "counting" did not appear in these papers. In fact, the research on discrimination between reinforcement schedules emerged *sui generis* from the tradition of Skinnerian behaviorism without connecting with any of the preceding literature on counting in animals. In a contemporary chapter on the representation of number, Gallistel (1990) reinterpreted these experiments (Rilling, 1967) as evidence for counting and drew three novel interpretations from the data: (a) Sometimes animals may prefer counting to timing; (b) animals count proficiently using numbers larger than implied by the "one-two-three-many" hypothesis; and (c) animals count events rapidly, because the birds were responding at a relatively high rate.

Rats Count Shocks

The discovery that rats count the number of shocks is a second example of the emergence of a procedure for studying counting that emerged unintentionally from the animal learning laboratory (Seligman & Meyer, 1970). Consider a preparation in which a rat is reinforced with food on a variable interval schedule. Responding is completely suppressed when a variable number of unpredictable electric shocks are presented while the rat is lever pressing for food. The safety-signal hypothesis predicts that responding for food will resume when the animal discriminates that no more shocks will occur. Seligman and Meyer (1970) presented rats with three shocks in a predictable sequence and found that lever pressing resumed after the third shock. They interpreted these data to mean that the rats behaved as if they were able to "count to three." Once again, the purpose of this research program was not an experimental investigation of counting. A counting interpretation emerged when demanded as an explanation of the data. Seligman was one of the first learning theorists to employ cognitive constructs as interpretations of learning experiments. His use of the word *counting* emerged from a mind "prepared" for cognitive constructs.

Davis emerged as one of the major contemporary investigators of counting in animals. His credits include two excellent reviews (Davis & Memmott, 1982; Davis & Pérusse, 1988) and a programmatic research program. Like Rilling, Davis emerged from the environment of radical behaviorism that prevailed at the University of Maryland in the mid-1960s. Joseph Brady, who popularized the conditioned suppression procedure as a tool for the study of learned fear in rats, served as a member of Davis's dissertation committee. His dissertation (Davis & McIntire, 1969) examined the effects of noncontingent shock on conditioned suppression. It was written in the tradition of radical behaviorism and contained no hint of counting as a possible interpretation of the data. As Davis continued his research on the effects of noncontingent shock on conditioned suppression, counting emerged as a theoretical possibility (Davis, Memmott, & Hurwitz, 1975). Soon he joined Seligman in concluding that the effects of noncontingent shocks on conditioned suppression could be explained on the assumption that animals counted shocks, and counting crystallized as his major research interest.

Counting Rewards and Nonrewards

A third preparation for studying counting emerged from Capaldi's research on serial order instrumental learning in rats. If animals counted shocks, then it followed that they might also count food-reinforcers. Unlike Rilling and Davis who started their careers as Skinnerian behaviorists, Capaldi inherited the Hull–Spence tradition of learning theory in which theoretical constructs, for example, memory for preceding events, were valued. As early as 1964, Capaldi assumed that instrumental behavior on a particular trial was determined by the pattern of reinforcement and nonreinforcement on preceding trials (Capaldi, 1964, 1967). A subsequent paper (Capaldi & Verry, 1981) systematically varied the number of nonrewarded events in a series and found that the rats remembered how many nonrewarded events had occurred. These data led Capaldi and Verry (1981) to conclude, for the first time, that the rats employed some form of counting mechanism. By 1988, the word *counting* appeared in the title of Capaldi's papers and the authors (Capaldi & Miller, 1988a, 1988b) were forcefully concluding that rats automatically count reinforcing events. Unlike Koehler, who started with the goal of investigating counting in animals, Capaldi's conversion to the counting hypothesis was gradual and emerged from over 30 years of careful experimentation during which alternative explanations, such as timing or odor discrimination, were carefully considered and rejected.

Comparative Psychologists Return to Counting

A fourth and final example of the contemporary rediscovery of counting is the reemergence of interest in the problem of counting by comparative psychologists. Here the trigger for the renewal was the animal-language problem, so once again a research community came to counting indirectly. After interest in the animal-language question waned, some investigators turned to the question of counting in animals. According to the rationale for this research, because human counting depends on the use of words, animals should first learn a language as a prerequisite for a test of counting. Ferster's (1964) and Ferster and Hammer's (1966) research program on teaching arithmetic to chimpanzees, described earlier, was based on this assumption. Pepperberg's (1987) research provided a contemporary example of an approach to counting via language. An African grey parrot was first trained to use approximations to English in order to label collections of two to six objects when requested by the trainer. This labeling ability generalized to novel objects. The renewal of interest in the problem of animal counting by comparative psychologists brings the problem back to the roots of its historical origin.

This section examines the research programs of Capaldi, Davis, and Rilling as three exemplars of programmatic research on counting within the tradition of U.S. learning theory. Pepperberg's research was considered as an example of comparative psychology. In each case counting emerged as an interpretation of data obtained from a research program that initially had nothing at all to do with counting. Davis and Rilling initially worked within a tradition of Skinnerian behaviorism, whereas Capaldi inherited the Hull–Spence tradition of instrumental learning in the runway. Pepperberg's research emerged from the animal-language tradition. With the exception of Pepperberg, none of the original papers by these authors cited any references to Koehler's research program on counting. Darwin, Romanes, and the comparative study of intelligence did not provide a framework for these research programs. These investigators built on a foundation of research in animal learning that provided well-developed procedures for obtaining high-quality data from laboratory animals. These investigators arrived at a counting interpretation slowly, after a series of preceding papers eliminated competing explanations. In the case of my own data, the counting interpretation was provided over 20 years later by Gallistel (1989). The grand philosophical arguments about paradigm shifts appear to have had little impact on these laboratory investigators who left their colleagues a legacy of valid procedures for studying an intriguing problem in the laboratory.

COUNTING THE REVIEWS

Over the years, the topic of counting in animals has been reviewed extensively and at frequent intervals. Like the fast-forward control for a videocassette, the conclusions from these reviews highlight how the critics interpreted the accumulating knowledge. Usually the reviewers were scientific skeptics whose criteria for experimental excellence were always slightly ahead of the state-of-the-experimental art. They increased the sophistication of the definition of counting from a simple discrimination to a concept of number (Davis & Pérusse, 1989) and arithmetic operations such as subtraction and division on the representation of numbers (Gallistel, 1989). The tension between experimenters and reviewers produced a constantly rising standard of experimental excellence.

Bierens De Haan (1936) concluded that the evidence showed that animals cannot count. He criticized that experiments that claimed evidence for counting confounded time, "rhythm" in his words, with the counting process. When he replicated an experiment by Gallis (1932) by varying the interval between tests from 10 seconds to between 20 and 30 seconds, he found that errors in the counting discrimination increased. Spence continued Yerkes's work on the counting problem and inherited some of the chimpanzees previously used by Yerkes. In a review, Spence (1937) concluded that Gallis had successfully demonstrated that primates understand the concept of the numbers two and three. Apparently Spence missed the criticism of this study by De Hann.

The first comprehensive review solely devoted to the question of the number concept in animals is that of Honigman (1942). He began with the papers of Kinnaman (1902) and Porter (1904) and concluded with seven pages devoted to an extensive discussion of Koehler's work. He pointed out that Koehler had eliminated figural and positional cues from the simultaneous procedure and "rhythm" from the successive procedure. In his summary, Honigman gave Koehler credit for establishing two basic discriminations based on number, but ended his summary with the conclusion that a number conception in the human sense does not exist for animals. Honigman's modest conclusions were too much for Salman (1943), who savaged Honigman's review of the preceding year. Salman defined counting from the human standpoint as the capacity to transfer from one domain to another. He was able to show that none of the experiments on animals met this impossibly high criterion. He criticized

1. INVISIBLE COUNTING ANIMALS   31

Koehler's work because it failed to control for a long list of potential stimuli other than number. But Salman mainly criticized the artificiality of the training procedures, which he considered abnormal because of the extensive training required to demonstrate counting. His suggestion was to study counting as it normally occurs in wildlife, for example, birds' egg-counting behavior. Salman's real target was the experimental analysis of counting in the laboratory. He regarded the data set as a collection of "meaningless artifacts" (Salman, 1943, p. 216). Such tension between field and laboratory investigators is probably an unresolvable characteristic of the field of animal behavior.

Wesley (1961) arrived on the scene just as postwar knowledge about counting in animals was beginning to accumulate. Wesley's paper had the flavor of Watsonian behaviorism. Alleged extraneous cues were preferred to a cognitive interpretation of the number concept, and he was hypercritical of the research from Koehler's laboratory. He also criticized Arndt's experiment on the counting of peas by pigeons with the assumption that the discrimination was based on visceral feedback from the weight of food in the animal's stomach. He failed to suggest that this variable could be controlled by varying the magnitude of reward, and assumed that a discrimination of visceral feedback was somehow simpler than a discrimination based on number. Schieman's (1939) work could be discounted because of a failure to control for odor cues. After eliminating most of the experiments as inadequate, Wesley was left with the conclusion that monkeys count to three and no more.

Davis and Memmott's (1982) watershed review ushered in the contemporary era. With respect to Koehler, Davis and Memmott (1982) concluded, "Nevertheless, there is a strong suggestion that Koehler was successful in teaching several species of birds, including pigeons, budgerigars, ravens, parrots, and magpies, to count" (p. 551). Why is Davis and Memmott's conclusion different than Wesley's, which appeared in the same journal, *The Psychological Bulletin*, 21 years earlier? The data base from Koehler's laboratory remained constant. The change was in the attitude of the reviewers. Davis and Memmott laid the ghost of Watsonian behaviorism to rest. The topic of counting behavior in animals had, at last, tiptoed into an eddy of the mainstream of experimental psychology. No longer could a critic dismiss the research on the basis of an alleged, hypothetical external stimulus that remained unmeasured. Davis and Memmott concluded that a wide variety of animals from birds to primates can learn to count.

The thesis of this chapter holds that theoretical diversity is a hallmark of normal science. *Behavioral and Brain Sciences* is a journal of reviews with the unique feature of open peer commentary. David and Pérusse (1988) provided a review of counting in animals that finally brings the problem into the mainstream of animal behavior. They are basically conservative critics who raised the criterion for counting from acquisition of a number discrimination to transfer across experimental situations. The paradigm that all investigators of counting have in common is a commitment to study animal learning in a laboratory setting, but differences among investigators today remain as sharp as they have always been. For example, Davis and Pérusse assumed with Koehler that counting is "unnatural," whereas Capaldi and Miller (1988a, 1988b) and Gallistel (1988) assumed that counting is a characteristic of problem solving by animals in nature. Gallistel (1988), reversing Morgan, argued that counting is an extremely simple process, whereas Davis and Pérusse called counting a complex numerical skill. Some investigators hold that counting in humans and animals is basically similar, whereas others assume that the processes are basically different. For some philosophers paradigm unity among a research community is a Holy Grail. On the contrary, if members of a research community completely shared the same paradigm, scientists would become pedantic clones of each other. The commentaries following Davis and Pérusse's review demonstrate that paradigm diversity is associated with scientific creativity.

In the reviews of Gallistel (1989, 1990), we come full circle 100 years later, back to a view of counting very similar to the view that prevailed during the age of Darwin. Lindsay (1880) held out the possibility that animals had a "faculty of numeration," as he put it in the title of a chapter, which was similar to arithmetic. With Gallistel, we reach the faculty of division: number of events divided by time. Koehler was unable to liken the learned counting behavior observed in the laboratory with the behavior of the organism in nature. Gallistel provided a plausible link; successful foraging depends on a representation of number that allows for a process isomorphic to division and subtraction. At last a plausible phenomenon in nature has been identified for which numerical competence would have a selective advantage. Darwin and Romanes would have been pleased. Gallistel's call for research on the neural substrates of counting and timing echoes Gall's view from the early 19th century that counting and timing are faculties with a unique neural basis in the brain.

Finally, take a last look at Fig. 1.1 to consider the stage in which the invisible college of animal counting stands today. Stages 1 and 2 are now history. Stage 3 has arrived. In 1989, the cumulative record appears poised for further acceleration in the years to come. How soon Stage 4 will arrive depends on the ingenuity of my colleagues. I invite the reader to turn to their chapters in this book.

CONCLUSIONS

Three distinctly different research communities—comparative psychology, ethology, and learning theory—contributed experiments on counting in animals. The review begins with Gall, the phrenologist, who in 1810

1. INVISIBLE COUNTING ANIMALS

asked if animals are endowed with the faculty of numbers. Kinnaman, a comparative psychologist, conducted the first laboratory experiment 92 years later. Within a framework of Darwinian intelligence, the question was whether rhesus monkeys could learn to count. Kinnaman's answer was negative. Other U.S. comparative psychologists led by Yerkes tried and failed to establish the phenomenon. Between the failures in the laboratory and the onslaught of Watsonian behaviorism, U.S. research on counting in animals dwindled to a trickle by World War II.

Koehler, a German ethologist, conducted a research program between 1926 and 1956 that definitively established counting in animals. This research program spanned 30 years and produced 17 primary research papers on the problem. Koehler was distinct among ethologists because he believed that counting in animals was a learned behavior that could be observed in the laboratory, but not in nature. The two capacities for counting that emerged from Koehler's research were based on simultaneous and successive procedures. The simultaneous research eliminated the confound with the form of the simultaneously presented objects, whereas the successive work eliminated the confound with time or "rhythm." Over the years a total of eight different species, including humans, were employed in the counting project. Koehler did not believe that his results constituted "true" counting because the animals did not show transfer from a trained number to a new number. Koehler's work has remained relatively invisible to the animal behavior research community because of U.S. ethnocentricity and a general prejudice from "Clever Hans" against the research problem.

World War II brought research on counting in animals virtually to a halt. Watson devised the research strategy of placing learning as the number one problem in comparative psychology. In the United States, comparative psychology almost disappeared as learning theory became a dominant field. After World War II, the problem of counting in animals emerged by accident as a theoretical construct necessary to explain the data from experiments on traditional problems: schedules of reinforcement, conditioned suppression, and the partial reinforcement extinction effect. Eventually experimenters turned away from the traditional problems of learning theory to focus directly on the counting problem.

A cumulative record of the number of publications in counting between 1902 and 1988 reveals that only 103 papers have been published on this topic. The small number of papers and the fact that the definition of the problem has remained unchanged are unusual, because in the field of animal behavior fads often come and go within a span of 10 years.

Historical analysis fails to support a model of revolutionary change of the type proposed by Kuhn (1962) as an explanation for the rate of appearance of publications on counting in animals during the last 90 years. During this period of time the subject matter of psychology has changed three times: from consciousness to behavior to cognition. These revolutions had relatively little impact on the research on counting in the laboratories.

Three invisible colleges—comparative psychology, ethology, and learning theory—have produced all of the experiments on counting. One of the three colleges dominated research in each decade under consideration but none achieved ultimate hegemony. Members of these colleges shared a paradigm of studying learned behavior in the laboratory, but they disagreed about practically everything else. Although many commentators long for a coherent paradigm among scientists, the thesis here purports that paradigm diversity is a causal factor in fertile periods of scientific productivity. If all scientists thought alike, they would run the same experiments.

The behavior of the scientists who contributed to counting in animals provides strong support for Richards's (1987) natural selection model. The intellectual environments and traditions of the scientists who contributed to the problem were quite different. From this theoretical diversity the scientists selected questions that were answered affirmatively by original experiments. During some decades progress was more rapid than in others, but progress was never radically discontinuous.

The problem of counting in animals has remained controversial because of the risk of anthropomorphism. Throughout the history of the problem, scientific critics have questioned the validity of the data from experiments. First they pointed to alternative sources of stimulus control. Then, with confounded sources of stimulus control eliminated, they called for transfer on the first trial from a trained to a novel number. The most recent demand is that of Davis and Pérusse (1988), who called for the demonstration of a concept of number established by the capacity to transfer across sense modalities. By setting a criterion slightly ahead of the data from the laboratories, the critics cause frustration in experimenters, yet they also help to raise the standard of excellence from the laboratory.

Counting in animals has remained invisible in textbooks on animal behavior. The tale of animal counting touches major issues in behavior theory and has now been told, so perhaps the writers of textbooks will soon rescue the problem from its past and undeserved obscurity.

ACKNOWLEDGMENTS

This research was supported by a grant from the Michigan State University Foundation. I thank Lauren Harris for his historical perspective and constructive comments on this chapter.

REFERENCES

Arndt, W. (1939). Abschließende Versuche zür Frage des "Zähl"—Vermögens der Haustaube [Concluding research on the question of "number" ability in the domestic pigeon]. *Zeitschrift für Tierpsychologie, 3*, 88–142.
Bierens de Haan, J. A. (1936). Notion du nombre et faculte de counter chez animaux [The idea of number and the faculty of counting in animals]. *Journal de Psychologie: Normale et Pathologique, 33*, 373–413.
Boakes, R. (1984). *From Darwin to behaviourism: Psychology and the mind of animals.* Cambridge, England: Cambridge University Press.
Capaldi, E. J. (1964). Effect of N-length, number of different N-lengths, and number of reinforcements on resistance to extinction. *Journal of Experimental Psychology, 68*, 230–239.
Capaldi, E. J., & Verry, D. R. (1981). Serial order anticipation learning in rats: Memory for multiple hedonic events. *Animal Learning and Behavior, 9*, 441–453.
Capaldi, E. J., & Miller, D. J. (1988a). Counting in rats: Its functional significance and the independent cognitive processes which comprise it. *Journal of Experimental Psychology: Animal Behavior Processes, 14*, 3–17.
Capaldi, E. J., & Miller, D. J. (1988b). Number tags applied by rats to reinforcers are general and exert powerful control over responding. *Quarterly Journal of Experimental Psychology, 40B*, 279–297.
Coburn, C. A., & Yerkes, R. M. (1915). A study of the behaviour of the crow by the multiple choice method. *Journal of Animal Behavior, 5*, 75–114.
Crane, D. (1972). *Invisible colleges: Diffusion of knowledge in scientific communities.* Chicago: University of Chicago Press.
Davis, H., & McIntire, R. W. (1969). Conditioned suppression under positive, negative, or no contingency between conditioned and unconditioned stimuli. *Journal of the Experimental Analysis of Behavior, 12*, 633–640.
Davis, H., Memmott, J., & Harwitz, H. M. B. (1975). Auto contingencies: A model for subtle behavioral control. *Journal of Experimental Psychology: General, 104*, 169–188.
Davis, H., & Pérusse, R. (1988). Numerical competence in animals: Definitional issues, current evidence, and a new research agenda. *Behavioral and Brain Sciences, 11*, 561–615.
Davis, H., & Memmott, J. (1982). Counting behavior in animals: A critical evaluation. *Psychological Bulletin, 92*, 547–571.
Emmerton, J. (1988, November). *Generalized numerosity discrimination and seriation by pigeons.* Paper presented at the meeting of the Psychonomic Society, Chicago, IL.
Ferster, C. B. (1958). Intermittent reinforcement of a complex response in the chimpanzee. *Journal of the Experimental Analysis of Behavior, 1*, 163–165.
Ferster, C. B. (1964). Arithmetic behavior in chimpanzees. *Scientific American, 210*, 98–106.
Ferster, C. B., & Skinner, B. F. (1957). *Schedules of reinforcement.* New York: Appleton-Century-Crofts.
Ferster, C. B., & Hammer, C. E. (1966). Synthesizing the components of arithmetic behavior. In W.K. Honig (Ed.), *Operant behavior: Areas of research and application* (pp. 634–676). New York: Appleton-Century-Crofts.
Fischel, W. (1926). Haben Vogel ein "zahlengedächtnis" [Do birds have a memory for numbers]? *Zeitschrift vergleichende Physiologie, 4*, 345–369.
Gall, F. J. (1835). *The influence of the brain on the form of the head* (W. Lewis Jr., Trans.). Boston: March, Capen, & Lyon. (Original work published 1810)
Gallis, P. (1932). Les animaux savent-ils compter [Can animals count]? *Bulletin of Social Science Liege*, 82–84.
Gallistel, C. R. (1988). Counting versus subitizing versus the sense of number. *Behavioral and Brain Sciences, 11*, 585–586.

Gallistel, C. R. (1989). Animal cognition: The representation of space, time, and number. *Annual Review of Psychology, 40*, 155–189.
Gallistel, C. R. (1990). *The organization of learning.* Cambridge, MA: Bradford Books, MIT Press.
Hassenstein, B. (1974). Otto Koehler—Sein Leben und sein Werk [Otto Koehler—His life and his work]. *Zeitschrift für Tierpsychologie, 35*, 449–464.
Hediger, H. K. P. (1981). The Clever Hans phenomenon from an animal psychologist's point of view. In T. A. Seboek & R. Rosenthal (Eds.), *The Clever Hans phenomenon: Communication with horses, whales, apes, and people* (Annals of the New York Academy of Sciences, Vol. 364). New York: New York Academy of Sciences.
Honigman, H. (1942). The number conception in animal psychology. *Biological Review, 17*, 315–337.
Hunter, W. S. (1913). The delayed reaction in animals and children. *Behavior Monographs, 2*(6).
Kinnaman, A. T. (1902). Mental life of two Macacus rhesus monkeys in captivity. II. *American Journal of Psychology, 13*, 173–218.
Koehler, O. (1941). Vom Erlernen unbenannter Anzahlen bei Vögeln [Learning about unnamed numbers by birds]. *Die Naturwissenschaften, 29*, 201–218.
Koehler, O. (1943). "Zähl"-versuche an einem Kolkraben und Vergleichsversuche an Menschen ["Number" ability in a raven and comparative research with people]. *Zeitschrift für Tierpsychologie, 5*, 575–712.
Koehler, O. (1951). The ability of birds to count. *Bulletin of Animal Behaviour, 9*, 41–45.
Koehler, O. (1957). *Thinking without words.* Proceedings of the 14th International Congress of Zoology, Copenhagen.
Köhler, W. (1925). *The mentality of apes.* New York: Liveright.
Kuhn, T. S. (1962). *The structure of scientific revolutions.* Chicago, IL: University of Chicago Press.
Lindsay, W. L. (1880). *Mind in the lower animals in health and disease.* New York: Appleton.
Morgan, C. L. (1895). *An introduction to comparative psychology.* London: Walter Scott.
Mountjoy, P. T., & Lewandowski, A. G. (1984). The dancing horse, a learned pig, and muscle twitches. *Psychological Record, 34*, 25–38.
O'Donnell, J. M. (1985). *The origins of behaviorism: American psychology* (pp. 1870–1920). New York: New York University Press.
Pepperberg, I. M. (1987). Evidence for conceptual quantitative abilities in the African Grey Parrot: Labelling of cardinal sets. *Ethology, 75*, 37–61.
Popper, K. (1968). *Conjectures and refutations: The growth of scientific knowledge* (2nd ed.). New York: Harper & Row.
Porter, J. P. (1904). A preliminary study of the psychology of the English sparrow. *American Journal of Psychology, 15*, 313–346.
Richards, R. J. (1987). *Darwin and the emergence of evolutionary theories of mind and behavior.* Chicago: University of Chicago Press.
Rilling, M. E. (1967). Number of responses as a stimulus in fixed-interval and fixed-ratio schedules. *Journal of Comparative and Physiological Psychology, 63*, 60–65.
Rilling, M., & McDiarmid, C. (1965). Signal detection in fixed-rate schedule. *Science, 148*, 526–527.
Romanes, G. J. (1883). *Animal intelligence.* New York: Appleton.
Romanes, G. J. (1898). *Mental evolution in man: Origin of human faculty.* New York: Appleton.
Rosenthal, R. (1965). Clever Hans. (Carl L. Rahn, Trans.). New York: Holt, Rinehart & Winston. (Original work published 1911)

Salman, D. H. (1943). Note on the number conception in animal psychology. *British Journal of Psychology, 33*, 209–219.
Samuelson, F. (1985). Organizing for the kingdom of behavior: Academic battles and organizational policies in the twenties. *Journal of the History of the Behavioral Sciences, 21*, 33–47.
Schiemann, K. (1939). Vom Erlernen unbenannter Anzahlen bei Dohlen [Learning about unnamed numbers by jackdaws]. *Zeitschrift für Tierpsychologie, 3*, 292–347.
Seboek, T. A., & Rosenthal, R. (1981). *The Clever Hans phenomenon: Communication with horses, whales, apes, and people* (Annals of the New York Academy of Sciences, Vol. 364). New York: New York Academy of Sciences.
Seligman, M. E. P., & Meyer, B. (1970). Chronic fear and ulceration in rats as a function of unpredictability of safety. *Journal of Comparative and Physiological Psychology, 73*, 202–207.
Skinner, B. F. (1965). Stimulus generalization in an operant: A historical note. In D. I. Mostofsky (Ed.), *Stimulus generalization* (pp. 193–209). Stanford, CA: Stanford University Press.
Spence, K. (1937). Experimental studies of learning and the higher mental processes in infrahuman primates. *Psychological Bulletin, 34*, 806–850.
Swets, J. A. (1961). Is there a sensory threshold? *Science, 134*, 168–177.
Thorndike, E. L. (1898). Animal intelligence: An experimental study of the associative processes in animals (Monograph Supplement No. 8). *Psychological Review*, 68–72.
Tinbergen, N. (1963). On aims and methods of ethology. *Ethology, 20*, 410–433.
Watson, J. B. (1914). *Behavior: An introduction to comparative psychology*. New York: Holt.
Wesley, F. (1961). The number concept: A phylogenetic review. *Psychological Bulletin, 58*, 420–428.
Yerkes, R. M. (1911). *Introduction to psychology*. New York: Holt.
Yerkes, R. M. (1916). The mental life of monkeys and apes: A study of creational behavior. *Behavior Monographs, 3*, 1–144.
Yerkes, R. M. (1934). Modes of behavioural adaptation in chimpanzees to multiple-choice problems. *Comparative Psychology Monographs, 10*, 1–108.
Yerkes, R. M., & Coburn, C. A. (1915). A study of the behaviour of pig (*sus scrofa*) by the multiple choice method. *Journal of Animal Behavior, 5*, 185–225.

CHAPTER TWO

Counting in Chimpanzees: Nonhuman Principles and Emergent Properties of Number

Sarah T. Boysen
Ohio State University
and
Yerkes Regional Primate Research Center
Emory University

Despite historically bad press, the study of numerical abilities or capacities in nonhuman species is currently provoking renewed interest and controversy among those with interests in animal cognition (Davis & Pérusse, 1988; Rilling, chap. 1, this vol.). Indeed, this very volume, devoted to studies of number-related skills among nonhuman and human species, attests to the revitalization of the field. The surrounding chapters highlight the myriad of questions that remain to be addressed from the current data, and suggest new questions that may be explored as pivotal issues related to animal counting come more sharply into focus. Although sensitivity to quantity has been demonstrated among numerous species (Capaldi & Miller, 1988; Davis, 1984; Hicks, 1956; Matsuzawa, 1985; Pepperberg, 1987; Rumbaugh, Savage-Rumbaugh, & Hegel, 1987; Woodruff & Premack, 1981), evidence for counting that is more analogous to the accomplishments of young children may be limited to recent work with chimpanzees (Boysen, 1992a; Boysen & Berntson, 1989; Thomas & Lorden, chap. 6, this vol.).

However, the establishment of such skills in a species even as intelligent as the chimpanzee requires a heroic effort (Boysen, in press), when compared to the seeming ease with which counting skills are acquired by preschoolers. There is little question that such a process is quite complex and unfolds over the course of years in human children (Fuson, 1988; Gelman & Gallistel, 1978). Similarly, our research on the emergence of counting in chimpanzees suggests that the acquisition of counting abilities is a complex process for this species as well, and also unfolds over years, not weeks or months (Boysen, 1992a). As an animal model, the chimpanzee is an apt student for exploring numerical competence for several reasons. First, the chimpanzee is a highly social ape who can be engaged in interactive teaching situations with its human teachers (Boysen, 1992b; Oden & Thompson, 1992). Second, previous work in a variety of laboratories exploring artificial language systems (Fouts, Fouts, & Schoenfeld, 1984; Gardner, Gardner, & van Cantfort, 1989; Matsuzawa, 1985; Premack, 1986; Rumbaugh, 1977; Savage-Rumbaugh, 1986) have clearly established the chimpanzee as a strong candidate for acquiring complex concepts, and it was with this background in mind that we began our studies of counting capacities in the chimpanzee.

Studies of number-related skills in chimpanzees are not unprecedented (Dooley & Gill, 1977; Ferster, 1964; Hayes & Nissen, 1971; Matsuzawa, 1985). In addition, Menzel (1960) reported what appeared to be a natural propensity in the chimpanzee to evaluate food portions, which would seemingly require some level of quantitative judgment of size and/or proportion. Similar tendencies have been observed in other captive chimps during acquired food-sharing, where choices of food shared by the animals appeared often to be a function of size or volume, with the choosing animal reliably selecting the largest portion or the cup containing the larger volume (Boysen, personal observations).

Whereas the task sequence and training on number concepts have been reported in detail elsewhere (Boysen, 1992a; Boysen & Berntson, 1989; Boysen & Berntson, 1990), an overview of the chimps' training would be useful, prior to discussion of more recently completed studies. When we first began exploring numerosity, the three chimpanzees in the lab (Darrell, age 4½; Kermit, age 4; and Sheba, age 3 at the time), spent virtually all day in the company of their human teachers, with brief periods of work on numerous cognitive tasks (sorting or matching colors and shapes; learning to recognize body parts—e.g., pointing to their ear when their teacher pointed to hers; drawing; matching photographs of foods to real foods, etc.). Such tasks were completed in short sessions of approximately 15–20 minutes, interspersed throughout the day, with the balance of the day with teachers spent playing chase games, tickling, taking walks outside the laboratory, and similar activities. In this setting, the first number-related task began as a daily game. The chimps were to place individual pieces of monkey biscuits in each of six compartments of a partitioned tray. The idea was to demonstrate one-to-one correspondence between the number of compartments and the biscuits, because one and only one biscuit was to be placed in each section. We elected to use monkey biscuits because they were readily available, not considered highly

2. COUNTING IN CHIMPS   41

palatable by any of the chimps but more important, not really interesting enough to distract them from the task at hand. The animals were tested individually, during which they were given a small bowl containing 8–10 pieces of chow. They quickly learned the game, and were able to accomplish the task within several sessions.

The next numbers task was designed to elaborate on one-to-one correspondence, and provide a conceptual framework onto which Arabic numerals would later be mapped. In this task, the animals learned to match the number of markers affixed to different placards with the corresponding number of gumdrop candies presented to them (Fig. 2.1). All trials initially consisted of the presentation of single gumdrops, until the animals were able to track the various locations of the single marked placard. Next, trials that consisted of presentations of two gumdrops per trial were run, with the animals now required to track the location of the placard with two markers among the three possible locations (see Fig. 2.1). Following criterion, which required 90% correct response on two successive sessions, the animals then completed sessions with trials in which one or two gumdrops were presented. They were now required to make an explicit one-to-one match between the number of candies presented on a given trial, and the corresponding marked placard. After reaching criterion on this phase of training, a third marked placard was introduced, and the animals trained until criterion performance was met with 1, 2, or 3 candies presented on a given trial.

The transition from the one-to-one correspondence task with the marked placards and the association of Arabic numerals with specific

FIG. 2.1.   Training stimuli for One-to-One Correspondence Task and introduction of Arabic numerals.

quantities was accomplished by systematically substituting each marked placard with its respective numeral. As depicted in Fig. 2.1, the numerals were introduced one at a time, and the animals' performance permitted to stabilize at the 90% criterion level before introducing the next number. The Arabic numerals 1, 2, and 3 were introduced to the chimps in this fashion, whereas all subsequent numerals were introduced directly as Arabic numerals, without the marked placard training phase. This included numerals 0 and 4-8 for Sheba; 0 and 4-7 for Darrell, and 0-5 for Kermit and Sarah.

A number comprehension task was the next phase of training for the chimps, although procedural changes in the task had to be made almost immediately. We were interested in providing the animals with a task analogous to a receptive language task, reasoning that like other symbol systems, numbers would need to be manipulated in both productive and receptive modes in order to achieve representational status (Savage-Rumbaugh, 1986). Thus, although the animals were able to produce a label for different-sized arrays by pointing to the correct Arabic numeral, similar to naming an object as we do with spoken language, the chimps also needed to demonstrate that they understood the *meaning* of a number if it was used by someone else. This would be analogous to a person being able to label an object (i.e., "cup") when it was presented for naming (productive mode), and then being able to select the correct object (i.e., select the cup from among several objects) when asked, "Show me the cup" (receptive mode). Both modes of symbol use are necessary for any type of dialogue to unfold around a particular representational system, and thus numbers, as a symbolic, representational system, seemed to warrant establishment of both receptive and productive symbol use. Other studies with chimpanzees trained on an artificial, visual language system demonstrated that the animals could not automatically translate their productive labeling skills to the receptive mode, but rather, both types of symbol use had to be trained separately (Savage-Rumbaugh, Rumbaugh, & Boysen, 1980).

To train the animals on the receptive number comprehension task, we initially planned to present two different-sized arrays per trial, as the chimps viewed a videomonitor displaying a single numeral 1, 2, or 3. Thus, for example, if the number 3 was shown, the chimp was to point to the array consisting of three gumdrops, and ignore the array with one gumdrop. As in all number-related tasks, the animal always got to eat the "stimuli" after each correct response. In the receptive task, they were permitted to eat the array they chose, with the idea being they would learn to pick the array that matched the number presented on the monitor. The chimps did not necessarily see the task in the same way we did, as they appeared to completely ignore the number being presented, and began to consistently choose the larger array of the two being offered. After several sessions of bleak performance, with no indication that they could inhibit this response pattern, we elected to substitute the arrays of gumdrops with the original one-to-one correspondence placards (see Fig. 2.1). Now the chimps were to view the number on the monitor during each trial, and select from among the three placards which had one, two, or three markers (Fig. 2.2). This procedural change made a significant difference in their performance almost immediately. The animals were quickly able to learn to attend to the videomonitor, which they had previously ignored, and select the placard with the corresponding number of markers. Darrell's performance was slightly better than Kermit's or Sheba's in reaching criterion, but overall, all three chimps' receptive performance was quite similar (Darrell, 201 trials to criterion, 70% CR; Kermit, 315 trials to criterion, 72% CR; Sheba, 282 trials to criterion, 69% correct response [CR]).

These four structured numbers tasks—one-to-one correspondence, matching arrays to marked placards, the transition to matching arrays with the appropriate Arabic numeral, and the number comprehension task—evolved one from the other over two years of training. In an effort to broaden the chimps' use of number concepts, additional number tasks were devised for the youngest animal, Sheba. Sheba, by now age 6, was still quite tractable, and was able to move unrestrained throughout the laboratory during the day. Kermit and Darrell, however, had entered into the tumultuous period of chimpanzee male adolescence, and were not able to move about freely because of their large size. Today, at age 10, Sheba continues to work interactively with her teacher in the

FIG. 2.2. Task format and training stimuli for Receptive Number Task.

same manner throughout the day, and has shown no aggressive or noncompliant behavior. Although some of the difficulties of working with male chimps into adulthood are gender-related, some may be dispositional with respect to the individual animals' personalities and other early experiential influences. Kermit and Darrell continue to readily engage in tasks with their teacher through the front of their home cages on a daily basis, as they did throughout the 3- to 5-year period associated with adolescent changes in their physical size and demeanor. However, at the time the new number tasks were designed for Sheba that required free movement in the lab, we elected to focus our efforts on these tasks with this single female subject.

Interests centered around whether or not a chimpanzee, trained on the structured numbers tasks described earlier, could learn to "sum" arrays presented to her in a somewhat more naturalistic encounter. That is, if Sheba were to come upon an array of size X in the lab, and next locate a second array of size Y, could she come to learn to combine the two arrays, and report a total number representing the two arrays? This is precisely the situation we presented to her, and the results were surprising for several reasons. Three food sites were selected for use in the first task, called the *functional counting task*: (a) a tree stump formerly used as a tool site by the animals, located at the far end of the lab; (b) a stainless steel food bin attached to the front of an unused cage, located approximately 10 feet from the stump; and (c) a plastic dishpan, positioned on the floor, approximately 20 feet from the stump, such that the three sites formed a rough triangle (see Fig. 2.3). Sheba's Arabic numerals 0–4 were positioned in ordinal sequence on a wooden platform (work station) located between Site 1 (stump) and Site 3 (dishpan). Sheba was accustomed to working on most tasks while sitting on this low platform with her teacher (Fig. 2.3). With the three sites available, the teacher (or a second experimenter, as during testing) placed from 0–4 oranges in two of the three sites, such that the arrays of oranges were not directly visible by Sheba as she sat at the work station. Sheba's task was to move among the three sites, return to the work station, and select the Arabic numeral that represented the total number of oranges hidden among the three sites. It was startling that Sheba was able to demonstrate the ability to complete this task from the very first session, moreover, from the very first trial. Rather than having to spend months training her on the many novel parameters of the functional counting task, she was able to readily grasp the rule structure of the testing arrangement immediately, and provide the correct total number during session one and maintain comparable performance throughout the 2-week testing period (Boysen & Berntson, 1989).

To challenge her further, given her ability to successfully complete the functional task, we next substituted the arrays of oranges with replicas

2. COUNTING IN CHIMPS 45

FIG. 2.3. Functional and Symbolic Counting Task: Physical setting, three sites for object and symbol stimuli, and numerical choices.

of the Arabic numeral placards that Sheba used as response stimuli. Now, instead of finding three oranges at Site 2, and one orange at Site 3, for example, Sheba came on the numeral 3 and the numeral 1 at the sites. Nonetheless, Sheba readily generalized her very limited experience in the functional counting task to this new paradigm, which we called the *symbolic counting task*. Her performance from Session 1, using symbols instead of object arrays, was significantly above chance, and she continued with similar performance levels for the balance of the study (Boysen & Berntson, 1989). One of the most remarkable things about Sheba's performance on both the functional and symbolic counting tasks was the fact that these novel tasks involved individual task components entirely new to her. Nonetheless, she was able to spontaneously move beyond the structured counting tasks described earlier, and readily solve both novel tasks, which required summation and/or maintaining a running tally of two separate arrays.

Although initial assessment of Sheba's performance on these tasks seemed somewhat inexplicable, an examination of the literature on early addition with human children offered some interesting insights (Groen & Parkman, 1972). Children as young as 3 are able to utilize spontaneous addition algorithms to solve simple problems, long before they receive any formal training in arithmetic (Starkey & Gelman, 1982), with such abilities growing out of their early counting experience and emerging concept of number. Sheba's performance on the functional and symbolic counting tasks suggested that she, too, was using a spontaneous addition algorithm, although the specific process by which she was solving the task remains unknown. She may have been counting both arrays sequentially, maintaining a running count, until she got to the last item, or she may have tallied the first array, and begun the count of the second array with that cardinal number. Both strategies are used by children as addition strategies, and are known respectively as "counting-on" and "counting-all" (Fuson, 1982, 1988; Groen & Resnick, 1977). Whatever strategies she was using, it is clear that Sheba had more skills with numbers than could be accounted for by the individual tasks for which she had been trained. Some emergent properties associated with the counting process became available to her for application with sequential arrays the first time they were presented to her. In addition, Sheba's performance on the symbolic counting task, in particular, suggested that number symbols served as representations for her—objects and numerals were interchangeable in their use. Number symbols were thus abstract representations of real-world referents, and could be readily manipulated in a novel context by Sheba, to characterize quantities represented by object arrays or other Arabic symbols. Her ability to perform both new tasks at the first opportunity also suggested that some emergent properties of numbers that were readily available for these novel complex tasks may have been the by-products of more directed training on counting. Sheba's abilities were highly reminiscent of similar transitions in counting skills and number concepts in preschoolers, and the process whereby such skills are acquired in children is also currently not well understood.

MOTOR TAGGING DURING COUNTING

In addition to the emergence of counting principles that seemed to be "beyond the information given," Sheba also began to exhibit tagging behavior and other indicating acts (Fuson, 1988; Gelman & Gallistel, 1978) during counting. Such behaviors first began with the introduction of the number 4, which was the first Arabic symbol introduced directly as a number symbol. Recall that the chimps had learned the numbers 1–3 by first learning to match the number of items in an array with placards bearing one, two, or three markers (see Fig. 2.1). When the number 4 was introduced directly as an Arabic numeral, Sheba began to move the items to be counted to new locations, push them apart, touch them, tap them, or simply point to them, prior to making her choice of the corresponding Arabic numeral (Fig. 2.4). Although such behaviors were noted by her teachers at the time, and comments added to her daily records, regrettably, no filmed record of the emergence of these tagging behaviors was made. As the behavior became more refined and consistent, we hypothesized that motor tagging could be functioning in a similar capacity as it did for children in the early stages of learning to count. That is, tagging may have helped Sheba keep track of the items in an array that she had already counted, and those that remained to be counted (Fuson, 1988; Gelman & Gallistel, 1978). To evaluate her tagging performance, all counting sessions over a 3-month period were videotaped, and Sheba's tagging behaviors were tallied. Two naive observers evaluated the videotaped sessions on a trial-by-trial basis, recording the number of tagging behaviors exhibited by Sheba, the number of repeat tags she made per trial, the correct number of items in the array, and the Arabic numeral Sheba chose on each trial. Inter-rater reliability between the two observers was .97, and disputed trials were resolved by a third observer. The high intercorrelations among the four measures revealed a close relationship between the number of items, Sheba's tagging responses, and her ultimate number choices. This is consistent with the proposition that Sheba used tagging of items in an array to help keep track of items she had counted, which may have permitted her to make a more accurate count (Table 2.1). It has been suggested by other investigators interested in animal counting that some covert process that permitted the animal to keep track of the items might be operating. For example, Koehler (1950) noted that the birds in his study might have employed some type of "inner marks," and used these to "think unnamed numbers," and such a process allowed the animal to keep track of items already counted. However, Sheba's tagging behavior may provide the first definitive evidence that such marking behavior is closely linked with both the number of stimulus items and the response.

Another significant issue in the counting debate centers around the principle of ordinality. In order to be considered as a candidate for "true counting," Davis and Pérusse (1988) suggested that an organism must demonstrate an appreciation for the ordered relations among and between numbers. That is, to say that one is counting is to assume they understand (and can demonstrate) that 2 is larger than 1, 5 is more than 3, 6 is greater than 4, and so forth—that the set of numbers used to label arrays of size N has a stable sequence, and represents an ascending series of increasing size. With human adults and children, we can simply pose the question verbally (e.g., "Is 4 more than 3?"). With nonhuman species, the evaluation of ordinality becomes more challenging.

To explore ordinality in the chimpanzee, we first sought to replicate

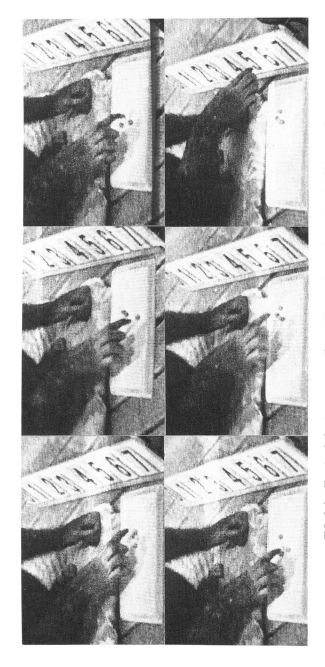

FIG. 2.4. Tagging behavior sequence (from videotaped counting sessions) with array of six items.

TABLE 2.1
Intercorrelation Matrix Among Variables
During Motor Tagging Task

	Sheba's Choice	No. of tags by Sheba	Items
Number of items	.73*	.77**	1.0
Number of tags by Sheba	.63***	1.0	
Sheba's choice	1.0		

*$t = 22.09, p < .0001$
**$t = 25.06, p < .0001$
***$t = 16.56, p < .0001$

Gillan's (1981) study of transitive inference in chimps. Transitive inference is an inferential judgment of the ordinal relationship between two elements, derived from premises that specify the relationship of each of the two elements to a third (Halford, 1984; Kingma & Zumbo, 1987). Typical test procedures with children consist of presenting a series of object pairs ABCDE (e.g., sticks of different colors and lengths), with each pair serving as a premise such that A > B > C > D > E. Under test conditions, the child is asked to identify the correct object of a nonadjacent pair, such as BD, which has not been explicitly trained. The assumption is that the child must derive the relationship between the two nonadjacent sticks, based on their relative position within the ordered series. The development of such tasks for use with young children, with minimal linguistic demands, led to models for testing of nonhuman primates, as in the Gillan (1981) study with chimpanzees and McGonigle and Chalmers (1977) with squirrel monkeys.

In a series of experiments similar to the earlier study of McGonigle and Chalmers (1977), Gillan (1981) used a series of colored boxes containing relative amount of foods, presented to young nonlanguage-trained chimpanzees. The boxes were presented serially in pairs, in the following sequence and respective contingencies: A – B +, B – C +, C – D +, D – E +. Following criterion performance on all nonadjacent pairs, the chimps were tested on the novel BD comparison. Two of the three animals tested eventually demonstrated consistent choice of D. From these and other tasks in the study, Gillan concluded that integration theories, suggesting that subjects integrate information about separate stimuli or pairs of stimuli into an ordered series, provided stronger support for his animals' data, and language training was not necessary to subserve use of transitive inference in the chimpanzee.

We viewed the Gillan (1981) procedure as a viable task for examining ordinality in our chimpanzees that had been previously trained in counting. Each of the chimps, however, evidenced considerable performance differences in their numerical abilities. At the beginning of the ordinality study, Sheba had the most breadth of experience with counting tasks, and also had the largest counting repertoire, from 0–6. Kermit and Darrell had begun training with numbers in identical fashion. However, early in his training, Kermit showed great difficulty in the transition to Arabic numerals, and continued to demonstrate confusion as each new number was added. Darrell, however, was able to perform consistently on numbers tasks with his Arabic numeral repertoire of 0–4, and was receiving concurrent training with the number 5 at the beginning of the ordinality study. Thus, despite their demonstrated individual differences in numerical abilities, all three chimps served as subjects in the ordinality study modeled directly after Gillan (1981).

The first phase of the training entailed an explicit replication of Gillan (1981), with the chimps trained with five pairs of colored boxes that formed an ascending, ordered series ABCDE. Boxes were always presented as adjacent pairs during training, with only one box of the pair baited. For instance, when Pair A – B + was presented, A never contained a food item, and B always did; for pair B – C + , B never contained food, and C always did, and so on. Thus, Box A was never baited, and Box E, when presented, always contained food, with these boxes representing the low and high endpoints of the box series. Pairs were trained in order, to a criterion of 90% or better for two consecutive, 16-trial sessions. Eventually, mixed-pair sessions were conducted, such that the animals had to reach criterion on each pair within a session. Following overall criterion performance, blind tests were completed in which probe trials with boxes BD were presented. All three chimpanzees selected Box D from the nonadjacent pair BD during novel testing, thus supporting the findings of Gillan (1981), that chimpanzees may be capable of employing transitive inference to determine the correct choice between two nonadjacent items in an ordered series (Boysen, 1988; Boysen, 1992a).

In a second experiment, in an effort to demonstrate an appreciation for ordinality in chimpanzees with experience in numerical tasks, Arabic numerals served as the training stimuli. Similar to the first task, adjacent number pairs between 1 and 5 were presented serially, and the chimps were to select the "larger" of the two symbols. If correct, the animals were reinforced with yogurt or juice, both nondiscrete edibles that would help avoid any task-specific association of the numerals with absolute numbers of reinforcers. The same 90% criterion was used, including mixed-pair sessions during which the animals had to reach criterion on all pairs in a given session. Novel tests were then conducted in which the nonadjacent, nontrained number pair 2,4 was presented as probe trials among randomly ordered training trials. Both Kermit and Darrell failed the first novel test, whereas Sheba consistently selected the number 4 when the novel 2, 4 pair was presented. These results suggested that Sheba recognized the ordinal relationships among the number series 1–5, and thus was able to report the larger number when a novel pair of numbers was presented. Although all three chimps had met the criterion performance demands of the training phase, both Kermit and Darrell were just as likely to select the number 2 as the number 4, when the novel 2,4 pair was presented. As seen in Table 2.2, both Kermit and Darrell had required a significantly greater number of trials to reach criterion in the final phases of training than were necessary for Sheba. Poor performance by Kermit was quite predictable, given his paltry performance with number concepts. Darrell's failure to pass the novel tests, however, was somewhat surprising, as he demonstrated a consistent ability to utilize Arabic numerals in both productive and receptive comprehension tasks, using numbers 0 through 4.

Sheba had both a larger counting repertoire than either Kermit or Darrell, as well as considerably more varied experiences with other number-related tasks, so this raised the question as to whether Sheba's success with both novel ordinality tests could be attributed to these training history differences. Perhaps the larger repertoire was a significant factor, and once Darrell (and presumably not Kermit, who still was not consistent in his use of 1–5) had expanded his counting ability to include additional numbers, he might be able to more readily represent the numbers as an ordered series. Sheba had also completed the reversal task, with training on number pairs in which she was now required to select the smaller number, and was successful during novel tests with the nonadjacent pair 2,4, so similar training was completed over several months with Kermit and Darrell. We reasoned that training with the descending pairs might help them further in organizing the numbers into a coherent series.

Following the descending series training ("smaller than") with Ker-

TABLE 2.2
Trials to Criterion for Number Pair Discriminations

Stimulus Pair	Sheba	Darrell	Kermit
Phase I: 1 vs. 2	111	12	27
Phase II: 2 vs. 3	71	56	226
Phase III: 3 vs. 4	36	104	155
Phase IV: 4 vs. 5	36	72	208
Phase V: 1 vs. 2, 2 vs. 3	72	44	158
Phase VI: 1 vs. 2, 2 vs. 3, 3 vs. 4	64	191	159
Phase VII: 1 vs. 2, 2 vs. 3, 3 vs. 4, 4 vs. 5	160	2,426	1,854

mit and Darrell, all three animals were given a refresher series of training sessions with ascending number pairs ("larger than"), approximately a year after initial testing. During the ensuing year, Darrell had received additional training with productive numbers, and had a fairly stable repertoire of 0 through 6. He also acquired facility with two fractions, $1/2$ and $1/4$, and had learned to create arrays of sizes 1, 2, 3, or 4 with wooden spools when Arabic numerals between 1 and 4 were presented. The latter task, known as Perceived Numbers, represented a high-level receptive skill that had emerged slowly by Darrell and Sheba over many months of training. However, different from most other number-related tasks completed in recent years, Darrell's performance on perceived numbers had exceeded that of Sheba's, whose progress was still painfully slow. Whether Darrell's conceptual breakthrough on the perceived numbers task reflected some new level of awareness relative to his sense of number remained to be seen.

The refresher sessions followed the same procedures as the original task, in which pairs of numbers were presented, and the chimps were to select the larger of the two. Following criterion performance on all four possible pairs (1,2; 2,3; 3,4; 4,5), probe trials in which the nonadjacent pair 2,4 were re-presented to the animals. Additional probe trials designated as "novel/novel," consisting of nonadjacent pairs of numbers that, with the exception of 1, were not part of the training stimuli, were also presented during novel testing. Sheba correctly selected the larger number of each probe pair, including all 12 trials of novel/novel stimuli. These novel/novel pairs included two trials each of 0,1; 4,6; 5,6; 5,7; 6,7; and 0,7. A total of 34 test trials were completed, including 20 training trials. Sheba's performance was 100% on the 2,4 trials, 100% on the novel/novel trials, and 90% on the training trials.

Both Kermit and Darrell also successfully completed the second novel tests, supporting the hypothesis that they had an appreciation for the ordinal features of the number series. Whereas Kermit's performance on the novel/novel probe trials did not reach significance, it should be emphasized that most of these numbers were completely unknown to him. All animals, however, were able to represent the numbers 1–5 in an ordered sequence, and both Sheba and Darrell were able to correctly select the larger number among pairs of nonadjacent, nontrained numbers. Darrell's data also suggests that familiarity with a greater number of Arabic representations, with which the animals had a working knowledge relative to their individual quantitative referents, contributed significantly toward organizing numbers into a coherent series.

CONSTRAINTS ON COUNTING CAPABILITIES?

Although some have required varying degrees of persistence in training, most number tasks that have been attempted with Darrell and Sheba had been learned by the animals. More recently, a study that involved a numerosity comparison task was designed to be conducted with two chimpanzees. Successful outcome of the task was dependent on the cooperation of both animals. The first phase of the study involved teaching the chimps a simple task in which they were required to compare two separate dishes of candies, each containing different numbers of food items. One chimp was supposed to select a dish, and the experimenter then would provide the contents of that dish to the second chimpanzee. The contents of the remaining dish would then be given to the chimp who had made the initial selection. Given the not-so-altruistic nature of chimpanzees when it comes to food, we had predicted that the chimps would eventually learn that if they picked the dish containing the smaller number of candies first, they would reap the larger amount from the second dish for themselves. Once the chimps learned this task, we proposed, we would introduce a cover for the dishes so that only one chimp would be able to see which dish had the larger amount. The chimpanzee who had visual access to the food dishes, however, would not have physical access to handles that operated a connecting mechanism between the two food dishes. If the chimp who knew where the food was located pointed to the correct dish, the other chimp, who had access to the rods, could pull on the correct rod, and bring the food dishes to within reach of each chimp. The chimps already knew how to operate the apparatus from an earlier study on perspective-taking (Povinelli, Nelson, & Boysen, 1990). For the comparison study described here, the goal was to see if the pointing chimp would attempt to deceive the other animal by indicating the dish with the smaller amount first. What began as a study of possible intentional deception between chimpanzees ended up as a very intriguing picture of possible perceptual preparedness, constraints on learning, and a glimpse at the power of symbols.

Sheba and Sarah were selected as subjects for the study, and Sheba was chosen to be the first selector chimp. The apparatus (Fig. 2.5) had originally had four pairs of food dishes, although only one dish from each pair was ultimately used in the present study. Two of the four dishes were baited with candies on each trial, with one dish containing one candy, and the other containing two pieces on a given trial. The selector chimp, Sheba, was free to choose either dish, with the rule that whatever dish was chosen first, its contents were given to Sarah. Eight sessions were

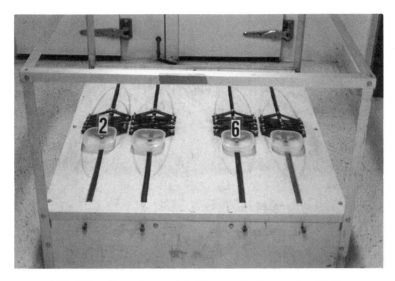

FIG. 2.5. Apparatus used in Numerosity Comparison Task.

completed with the 1:2 ratio, with Sheba failing to perform above chance. The ratio of candies was then increased to 1 versus 4, to make the difference between the two arrays more discriminable, and perhaps increase Sheba's motivation to learn the task. Chance performance was maintained for the next eight sessions, as well as during eight additional sessions with ratios of 1 versus 6 candies per trial. When the animals switched social roles, so that now Sarah was the selector, she fared no better than Sheba in learning the task. Here were two of the most highly trained chimpanzees in the world, unable to acquire what we had perceived as a simple discrimination problem.

Although Sarah was still in the early stages of learning Arabic numerals, recall that Sheba had an extensive counting repertoire that now included use of number symbols from zero to eight. Neither animal seemed to be able to learn to pick the smaller array of candies, so we abandoned hopes of studying deception, and instead sought to understand why the animals were unable to learn the comparison task. Both chimps seemed unable to inhibit selecting the larger of the two arrays, regardless of the ratio of candies (1 vs. 2; 1 vs. 4; 1 vs. 6). We hypothesized that the immediate presence of the reinforcers may be interfering with their ability to either acquire the underlying rule structure of the task, or the presence of the food was interfering with their ability to invoke the rule that would net them the greater reinforcement. To explore these issues, the food arrays were replaced with number placards

1 versus 2, 1 versus 4, and 1 versus 6. Sheba's performance from Trial 1, Session 1 was correct, and her overall performance for that session and the next two was significant. Thus, her performance on the comparison task when number symbols were used to represent the food arrays suggested that, in fact, she had learned the rules of the task. However, this latent knowledge was being expressed only when number symbols, and not the actual food arrays, were used as stimuli. On Day 4, arrays of edibles were reintroduced, and Sheba's performance plummeted to 17%. Her performance over the entire nine sessions of this testing phase is shown in Table 2.3. Without question, Sheba was able to demonstrate recognition of the rule structure of the task that would permit her to garner the larger number of candies. Yet, she appeared to only be able to express this knowledge if representational symbols were employed (Boysen & Raskin, 1990). Any time that actual food items were used, her performance deteriorated to well below chance. These results now provided an additional, new question: Why was Sheba able to solve the task when number symbols were used?

Recall that the ratios of the arrays always consisted of 1 versus 2, 4, or 6 candies, and thus the numbers used in the symbol sessions were 1 versus 2, 4, or 6. This raised the possibility that Sheba had fortuitously learned to select the numeral "1," and was continuing a "win-stay" strategy without necessarily recognizing the actual strategy for solving the problem (i.e., select smaller amount first; one receives larger remainder). To test this hypothesis, we ran eight test sessions in an ABBA design, during which edible arrays were used as stimuli for four sessions, and number combinations comprised of all possible numbers in Sheba's counting repertoire were used as stimuli in the other four sessions. Sheba's performance over the eight sessions was virtually identi-

TABLE 2.3
Numerosity Comparison Task: Alternating Sessions of
Numerical Comparisons and Food Arrays*

Test Stimuli		Correct Trials/Total	% CR
Day 1	Numbers 1 vs. 2, 1 vs. 4, 1 vs. 6	8/12	67
Day 2	Numbers	9/12	75
Day 3	Numbers	10/12	83
Day 4	Foods	2/12	17
Day 5	Numbers	9/12	75
Day 6	Numbers	8/12	67
Day 7	Foods	2/12	17
Day 8	Foods	2/12	17
Day 9	Numbers	9/12	75

*Sheba as selector

cal to the first test (Fig. 2.6). She was able to readily complete each novel symbol trial, reliably selecting the smaller number of the pairs. Thus, she consistently received the larger number of reinforcements on each trial, and Sarah got the smaller amount. These tests eliminated the hypothesis that Sheba had simply learned to pick the numeral "1" in the first test, when the symbolic comparison trials used only 1 compared with 2, 4, or 6. Rather, she was operating from a more general conceptual rule—"pick the smaller number first." These findings also suggested that Sheba had learned the task, even though she continued to be unable to express her knowledge of the conceptual rule if physical arrays were used. The use of numerical representations, however, permitted her to accurately and consistently decide between each symbolic comparison (Boysen & Raskin, 1990). In a very dramatic way, replacing the physical arrays with an abstract, arbitrary symbol freed Sheba from the limitations and constraints imposed by the immediate presence of the food items. If, however, her tendency to choose the larger food source is viewed as part of an innately endowed, adaptive mechanism that permits the rapid evaluation and/or estimation of quantity, it is not difficult to see how such a processing system would be advantageous for a foraging, visually oriented species like the chimpanzee. It is only within a cultural context such as a primate learning study, when the task at hand imposes a conceptual rule that, in principle, runs counter to a more adaptive response, that we were

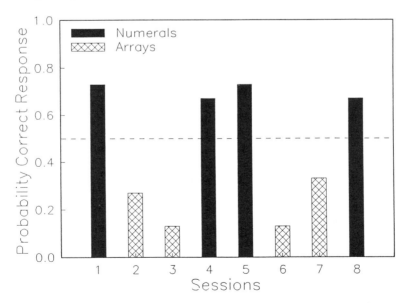

FIG. 2.6. Numerosity Comparison Task: Performance comparisons between novel number pairs and food arrays.

able to view the chimpanzee's inability to acquire the "correct" response. A myriad of questions remain with respect to the demonstrated performance differences with physical arrays and numerical representations, and the level of response preparedness for quantity judgments in the chimpanzee. Additional studies are currently underway in an effort to further clarify the relative perceptual and cognitive contributions to the animals' performance.

CONCLUSIONS

Studies of numerical competence in the chimpanzee continue to provide new insights into the range and capacity for quantitatively based information processing in this species. In general, the rebirth of studies of animal counting currently suggests that this area remains a rich and fruitful source of contributions to our understanding of animal cognition and behavior. And for a truly comparative perspective, it will be important for researchers to challenge their creativity, by continuing to devise new methods for tapping capacities toward counting in a variety of species, including nonhuman primates, rats, birds, and additional new species for whom no data currently exists.

REFERENCES

Boysen, S. T. (1992a). Counting as the chimpanzee views it. In G. Fetterman & W. K. Honig (Eds.), *Cognitive aspects of stimulus control* (pp. 367–383). Hillsdale, NJ: Lawrence Erlbaum Associates.
Boysen, S. T. (1992b). Pongid pedagogy: The contribution of human/chimpanzee interaction to the study of ape cognition. In H. Davis & D. Balfour (Eds.), *The inevitable bond: Examining scientist-animal interaction* (pp. 205–217). New York: Cambridge University Press.
Boysen, S. T. (1988, May). *Four is more: Evidence for an understanding of ordinality in chimpanzees (*Pan troglodytes*).* Paper presented at the annual meeting of the Midwestern Psychological Association, Chicago, IL.
Boysen, S. T., & Berntson, G. G. (1989). The development of numerical competence in the chimpanzee (*Pan troglodytes*). *Journal of Comparative Psychology, 103*, 23–31.
Boysen, S. T., & Berntson, G. G. (1990). The development of numerical skills in chimpanzees (*Pan troglodytes*). In S. T. Parker & K. R. Gibson (Eds.), *"Language" and intelligence in nonhuman primates* (pp. 435–450). New York: Cambridge University Press.
Boysen, S. T., & Raskin, L. S. (1990, November). *Symbolically-facilitated discrimination of quantities by chimpanzees.* Paper presented at the annual meetings of the Psychonomic Society, New Orleans, LA.
Capaldi, E. J., & Miller, D. J. (1988). Counting in rats: Its functional significance and the independent cognitive processes that constitute it. *Journal of Experimental Psychology: Animal Behavior Processes, 14*, 3–17.

Davis, H. (1984). Discrimination of the number "three" by a raccoon (*Procyon lotor*). *Animal Learning and Behavior, 12*, 409–413.

Davis, H., & Pérusse, R. (1988). Numerical competence in animals: Definitional issues, current evidence, and a new research agenda. *The Behavioral and Brain Sciences, 11*, 561–579.

Dooley, G. B., & Gill, T. (1977). Acquisition and use of mathematical skills by a linguistic chimpanzee. In D. M. Rumbaugh (Ed.), *Language learning by a chimpanzee: The Lana project* (pp. 247–260). New York: Academic Press.

Ferster, C. B. (1964). Arithmetic behavior in chimpanzees. *Scientific American, 210*, 98–106.

Fouts, R. S., Fouts, D. H., & Schoenfeld, D. (1984). Sign language conversational interactions between chimpanzees. *Sign Language Studies, 34*, 1–12.

Fuson, K. C. (1982). An analysis of the counting-on solution procedure in addition. In T. C. Carpenter, J. M. Moser, & T. A. Romberg (Eds.), *Addition and subtraction: A cognitive perspective* (pp. 67–81). Hillsdale, NJ: Lawrence Erlbaum Associates.

Fuson, K. C. (1988). *Children's counting and concepts of number*. New York: Springer-Verlag.

Gardner, R. A., Gardner, B. T., & van Cantfort, T. (1989). *Teaching sign language to chimpanzees*. Albany: State University of New York Press.

Gelman, R., & Gallistel, C. R. (1978). *The child's understanding of number*. Cambridge, MA: Harvard University Press.

Gillan, D. J. (1981). Reasoning in the chimpanzee: II. Transitive inference. *Journal of Experimental Psychology: Animal Behavior Processes, 7*, 150–164.

Groen, G. J., & Parkman, J. M. (1972). A chronometric analysis of simple addition. *Psychological Review, 79*, 329–342.

Groen, G. J., & Resnick, L. B. (1977). Can preschool children invent addition algorithms? *Journal of Educational Psychology, 69*, 645–652.

Halford, G. S. (1984). Can young children integrate premises in transitivity and serial order tasks? *Cognitive Psychology, 16*, 65–93.

Hayes, K. J., & Nissen, C. H. (1971). Higher mental functions of a home-raised chimpanzee. In A. M. Schrier & F. Stollnitz (Eds.), *Behavior of nonhuman primates* (Vol. 4, pp. 59–115). New York: Academic Press.

Hicks, L. H. (1956). An analysis of number-concept formation in the rhesus monkey. *Journal of Comparative and Physiological Psychology, 49*, 212–218.

Kingma, J., & Zumbo, B. (1987). Relationship between seriation, transitivity, and explicit ordinal number comprehension. *Perceptual and Motor Skills, 65*, 559–569.

Koehler, O. (1950). The ability of birds to "count." *Bulletin of Animal Behaviour, 9*, 41–45.

Matsuzawa, T. (1985). Use of numbers by a chimpanzee. *Nature, 315*, 57–59.

McGonigle, B., & Chalmers, M. (1977). Are monkeys logical? *Nature, 267*, 694–696.

Menzel, E. (1960). Selection of food by size in the chimpanzee, and comparison with human judgments. *Science, 131*, 1527–1528.

Oden, D., & Thompson, R. (1992). The role of social bonds in motivating chimpanzee cognition. In H. Davis & D. Balfour (Eds.), *The inevitable bond: Examining scientist-animal interaction*. New York: Cambridge University Press.

Pepperberg, I. M. (1987). Evidence for conceptual quantitative abilities in the African Grey Parrot: Labelling of cardinal sets. *Ethology, 75*, 37–61.

Povinelli, D. J., Nelson, K. E., & Boysen, S. T. (1990). Inferences about guessing and knowing by chimpanzees (*Pan troglodytes*). *Journal of Comparative Psychology, 104*, 203–210.

Premack, D. (1986). *Gavagai*. London: Cambridge University Press.

Rumbaugh, D. M. (1977). *Language learning in a chimpanzee: The Lana project*. New York: Academic Press.

Rumbaugh, D. M., Savage-Rumbaugh, E. S., & Hegel, M. (1987). Summation in the chimpanzee. *Journal of Experimental Psychology: Animal Behavior Processes, 13*, 107–115.

Savage-Rumbaugh, E. S. (1986). *Ape language: From conditioned response to symbol.* New York: Columbia University Press.

Savage-Rumbaugh, E. S., Rumbaugh, D. M., & Boysen, S. (1980). Do apes use language? *American Scientist, 68*, 49–61.

Starkey, P., & Gelman, R. (1982). The development of addition and subtraction abilities prior to formal schooling in arithmetic. In T. P. Carpenter, J. M. Moser, & T. A. Romberg (Eds.), *Addition and subtraction: A cognitive perspective* (pp. 99–116). Hillsdale, NJ: Lawrence Erlbaum Associates.

Woodruff, G., & Premack, D. (1981). Primitive mathematical concepts in the chimpanzee: Proportionality and numerosity. *Nature, 293*, 568–570.

CHAPTER THREE

Numerosity as a Dimension of Stimulus Control

Werner K. Honig
Dalhousie University

A concern with counting has dominated the study of discrimination of numbers in animals for many years (Capaldi & Miller, 1988; Church & Meck, 1984; Thomas, Fowlkes, & Vickery, 1980; Wesley, 1961; see Davis & Pérusse, 1988, together with commentaries, for a review). This attitude reflects an understandable preoccupation of adult humans with the domain of numbers. Counting is an advanced skill that requires categorical discriminations between different numbers, and a distinctive behavioral "tag" for each discriminated value. Counting presupposes a clear discrimination between adjacent numbers. With such stringent requirements, counting may be restricted to a small number of species, to particular circumstances, and (in subhuman animals) to small numbers of objects or events (Davis & Pérusse, 1988). For many animals, and perhaps young children, less precise discriminations of *numerosity* may be sufficient for the assessment of quantitative features of their environment. In this chapter I describe research that supports numerosity as a dimension of stimulus control, and I conclude that discrimination of values on this dimension is a process distinct from counting.

My terminology is similar to that of Davis and Pérusse (1988) in their review of numerical competence in animals. We have studied "relative numerousness" (which we call *relative numerosity*) with arrays composed of elements that differ in color, size, form, or conceptual category (Honig, 1991a, 1991b; Honig & Stewart, 1989). Pigeons readily discriminate this aspect of visual arrays of elements. In our research, the total numbers of elements in the arrays have ranged from 9 to 64. The discriminations are controlled by the proportions of the two different items, rather than their absolute number. The birds transfer the discrimination on a proportional basis among arrays that differ in the total numbers of elements (Honig, 1991a; Honig & Stewart, 1989). Therefore the birds judge the relative numerousness of the items in the array.

Our research indicates that the pigeons do not discriminate numerosity on the basis of some other, correlated variable, such as the sums of the areas of the different kinds of elements. They readily discriminate the relative numerosities of items that differ in form, such as X and O, and items that represent different conceptual categories, such as little pictures of birds and flowers. Such elements do not lend themselves to summation. Furthermore, they discriminate the relative numbers of elements that differ only in size (Honig & Stewart, 1989).

It is therefore most likely that the relative numerosities of the elements are discriminated directly. Even so, it would seem that such a discrimination must, at some level of perceptual analysis, involve the assessment of the absolute numbers of the different elements in the array. In the present research we studied the discrimination of absolute numerosity as an independent process. We worked mainly with small numbers of elements, as they are presumably easier to discriminate than large ones, and may be more important for the animal's commerce with its environment. For example, most birds and mammals raise small numbers of young in a litter. As becomes evident, that presupposition may not have been correct.

In this account I distinguish between numerosity discrimination, number discrimination, and counting. A *numerosity discrimination* involves a discrimination between nonadjacent numbers, or between different ranges of numbers of elements. A *number discrimination* involves differential responding to adjacent numbers of items, such as 3 and 4, 7 and 8, and so forth. *Counting* is a number discrimination in which different responses are made to each of a series of adjacent numbers of items. Although number discrimination is required for counting, it is not the same as counting. In a number discrimination, one response would be appropriate for the odd numbers of items, and a different response for the even numbers. *Counting* requires a unique behavioral indicator for each of several adjacent numbers. I conclude on the basis of data presented here, that pigeons are capable of numerosity discriminations, but number discriminations are more difficult, and performance is influenced by the method used for training. We have not worked with a procedure that would meet the present definition for counting.

DISCRIMINATION OF THE NUMBER OF PERSONS IN A PICTURE

Within their natural environment, animals might be expected to discriminate small numbers of natural objects. This experiment provides evidence for a numerosity discrimination with items presented in pictures. After the original study by Herrnstein and Loveland (1964), it has been shown many times that pigeons can discriminate the presence of one or more persons in a picture (see e.g. Edwards & Honig, 1987; Herrnstein, Loveland, & Cable, 1976; Vaughan & Herrnstein, 1987; Wright, Cook, Rivera, Sands, & Delius, 1988; and see Herrnstein, 1984; Lea 1984; and Wasserman & Blatt, 1992, for reviews.) If pigeons are also capable of numerosity discriminations, they should be able to discriminate among different numbers of people in a picture.

Method

Pigeons were trained in an operant behavior chamber to peck at a panel, 7.5 cm square, onto which pictures were back projected. The slides varied in content and appearance. Each slide displayed one or more persons. Two numerical categories were used in training: One person versus many persons, where "many" was defined as eight or more. Most of the slides represented outdoor scenes. People were shown at various distances from the camera, and in various positions and orientations. In many cases, one or more persons were partly obscured by other features, and in some cases, by other people.

The pigeons were trained with a successive operant discrimination procedure to discriminate between two sets of pictures. Sixteen slides with one person, and 16 slides with "many" persons, comprised the training sets. Each training session consisted of 32 trials. During each trial one slide was shown. A negative trial ended after 20 s. On a positive trial, the first response after 20 s provided 4 s of access to grain reinforcement. That last, reinforced response was not counted. Trials were separated by 10-s intertrial intervals, during which the screen was blank and responses had no effect.

After the pigeons learned this discrimination, they were tested for two sessions with several numerosity values. These sessions were run in extinction, and separated by one session of retraining. The test categories involved pictures with 1 person, with 2 persons, with 3 persons, with either 4 or 5 persons, with either 6 or 7 persons, and with "many" persons, defined as 8 or more. Four instances of each test category were presented in each session. The test slides with 1 person were all new, and so, obviously, were those of the test categories. Different instances were shown during the two test sessions. However, our entire collection of slides of 8 or more ("many") people had been used for training; therefore the training slides of "many" were also presented in the test procedure.

Results

The pigeons readily acquired the discrimination between one and many. Three birds reached a discrimination ration of .90 after 4 to 6 sessions. (This means that at least 90% of the total responses during a session were made to the set of positive stimuli.) One laggard required 12 sessions to reach the same criterion. The birds maintained the discrimination very well during the retraining phase between the two sets of test sessions.

The data from the two test sessions were similar, and have been combined. Figure 3.1 provides the data for each subject. Each bird provided a declining gradient, but the slopes of the gradients differed between subjects. The flattest gradient was obtained from a bird with "many" positive. The mean gradient has not been plotted in the figure, although it is very orderly. However, it obscures the marked differences in the slopes of the individual gradients.

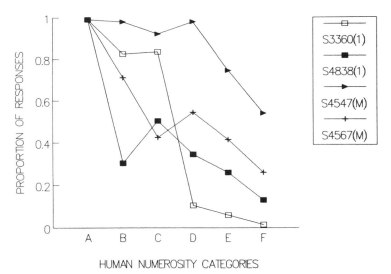

FIG. 3.1. Numerosity gradients obtained from four pigeons with different numbers of people in pictures. For S3360 and S4838 one person in the pictures was positive, and "many" (eight or more) were negative; the opposite contingency applied to S4547 and S4567. The numerosity categories are indicated as follows: A: 1 or 8+; B: 2 or 6-7; C: 3 or 4-5; D: 4-5 or 3; E: 6-7 or 2; F: 8+ or 1.

Discussion

This study shows that pigeons can discriminate one member of a natural category from several members when representations of the real world are used in training. The procedure establishes a continuum that generates orderly discrimination gradients on the dimension of number (or numerosity). However, in the casually assembled slides, the image of a single person shown on a slide was usually larger than the images of people in the slides that contained several persons. Although there was some overlap in size, the distributions of the sizes were markedly different. In order to determine whether this confounded difference controlled the pigeons' behavior, we sorted the sets of training pictures of one and of many persons into three categories: Those in which the person (or people) was large, medium, or small. The birds were trained for 12 further discrimination sessions as described earlier with the slides categorized accordingly. There was no evidence that this confound affected responding to the positive instances. The two pigeons for which one person was positive actually responded most to the slides in which the single person was small. There was some evidence of a confound with the negative instances. The two pigeons trained with one person positive responded more to the slides in which "many" were large, and one of the two birds with "many" positive responded most to negative slides in which the single person was small. However, in all cases the levels of responding to the negative slides were only about one tenth of those obtained with positive slides, and the differences in responding to these slides were not very reliable.

We were encouraged by this experiment to extend the discrimination among small numbers of items to arrays of abstract visual elements. The characteristics of such arrays could be more carefully controlled than the pictures of naturalistic scenes.

NUMBER DISCRIMINATION WITH VISUAL ELEMENTS

The experiment with pictures of natural scenes provided interesting and orderly results, but a systematic study of small number discriminations requires a method that permits a clearer definition of the stimulus elements, and a greater variety of patterns that are subject to experimental control and specification. To this end, we used patterns of small dots projected onto a response screen. The numbers of dots ranged from 2 to 7, but not all of these were used in every phase of the research. The dots were displayed on the response panel, 7.5 cm square, which was used for the discrimination of people in pictures, and the general method used for that research was in effect for these further experiments: Each trial lasted for at least 20 s and on positive trials, the first response following that interval was reinforced but not counted. All of the stimulus values involved in each discrimination were presented once in each block of trials. Several blocks of trials were run in each session; the numbers of blocks varied somewhat between experiments.

The total colored area of the items on the screen is necessarily correlated with their number. In order to eliminate this confounded variable, we used both small dots, 4 mm in diameter on the response screen, and large dots, 9.5 mm in diameter. The total colored area of the arrays of the smaller dots ranged from 25 sq mm for 2 dots, to 88 sq mm for 7 dots. For the large dots, this range extended from 113 sq mm to 397 sq mm. Thus the total area of the largest number of small dots was smaller than the total area of the smallest number of large dots. The dots were located at random in a square matrix that provided nine possible positions. With this arrangement, the mean distance between dots on the response screen was inversely correlated with their number. To compensate for this, we would have had to decrease the interitem distances in the less numerous arrays. That would have resulted in a different confounded cue: The total extent of the arrays would have been correlated with the number of dots in the arrays. Several patterns were prepared for each number of items. This reduced the confound between the number of elements and the interitem distance, but it could not eliminate it.

Procedure

The first experiment was a *number discrimination* involving 2, 3, 4, and 5 dots on the response screen. Each block of trials contained two patterns of dots representing each training value—one pattern of large dots and one of small dots. Each block therefore consisted of 8 trials, and 4 blocks comprised each training session. Two pigeons were trained with the arrays of 2 and 4 dots positive, and the arrays of 3 and 5 dots negative ("even numbers positive"). The other two were trained with the opposite contingency ("odd numbers positive"). The birds were trained for 12 blocks of 4 sessions each.

Results

The size of the elements had little effect: Differences in responding to large and small elements were negligible and not related in any way to the number of elements in the array. Therefore, the data obtained with the large and small elements have been combined for each numerosity value for each bird. These are presented in Fig. 3.2 from the 12th block of training sessions. S 8429, for whom 2 and 4 were positive, showed a slight declining gradient, which probably reflects the fact that on the

3. NUMEROSITY AS A DIMENSION OF STIMULUS CONTROL   67

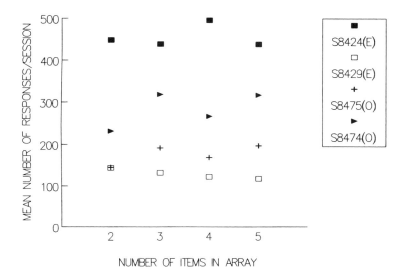

FIG. 3.2. The discrimination between arrays containing 2, 3, 4, or 5 stimulus elements. For S8424 and S8429, the even numbers (2 and 4) were positive. For S8475 and S8474, the odd numbers were positive.

average, the smaller numbers were positive. S 8424, again with even numbers positive, responded slightly more to the arrays of 2 and 4 dots than to the others. S 8476 and S 8475, for whom 3 and 5 dots were positive, responded differentially in accordance with the contingencies, but the discriminations are not impressive.

From the numbers of total responses we obtain a discrimination ratio (DR) that reflects performance. The ratio is obtained by dividing the responses to the positive stimuli by the total of positive and negative responses to all of the stimuli. A value of 1.00 reflects a perfect discrimination, whereas no discrimination between the stimuli results in a DR of .50. We obtained DRs for the individual birds by combining the data from the two positive number values, and from the two negative number values, for the eighth block of sessions. The results are follows: for S 8424 (2 and 4 positive), .51; S 8429 (2 and 4 positive), .51; S 8475 (3 and 5 positive), .53; S 2564 (3 and 5 positive), .55. Admittedly, this DR is not very sensitive to the initial stages of a discrimination; for example, if the responses to S+ exceed responses to S− by 50%, the DR is .60. Even so, the discriminations are not impressive.

In short, the birds do not seem capable of a number discrimination within the range of values used in this study. The two birds with 3 and 5 positive did show a modest discrimination of the contingencies. But even this was affected by a general trend reflecting the fact that, on the average, "more" was better. Likewise, *S* 8429 seemed to discriminate that on average, the smaller numbers were positive.

NUMEROSITY DISCRIMINATION WITH SMALL NUMBERS OF ELEMENTS

Aside from counting, the discrimination between adjacent numbers of elements is the limiting, and presumably the most difficult discrimination between integral numbers of elements. This number discrimination is required for counting, but animals that do not count may not benefit from such a fine discrimination. As we have seen, our pigeons achieved the number discrimination only to a modest degree. In three further experiments, we trained pigeons on a numerosity discrimination involving two ranges of small numbers—2, 3, 4, versus 5, 6, 7—in which one range was positive and the other was negative. This numerosity discrimination involved (by our definition) one number discrimination, between 4 and 5 elements. The most appropriate discrimination would be a clear distinction between the two ranges, and equal responding within each range. As pointed out here, the results were quite different.

In these three experiments, the discrimination between the two ranges of small numbers was the "final" discrimination. In Experiment 1, the pigeons were trained only on this discrimination. In Experiment 2, extensive training on the number discrimination between 4 and 5 elements preceded the final discrimination. In Experiment 3, we hoped to take advantage of the "easy-to-hard" effect. The pigeons were trained first with a discrimination between 2 and 7 elements. The other numbers of elements were then added in the final discrimination. This of course incorporated the number discrimination between 4 and 5 elements.

GENERAL PROCEDURES

The stimuli and the general training procedure were the same as in the previous experiment, except that the patterns of 6 and 7 dots were also used. As before, a negative trial ended after 20 s without reinforcement, whereas a positive trial ended with the first peck at the panel after 20 s, which was not counted, but reinforced. The different numbers of dots were presented twice in each block of trials—2, 3, 4, 5, 6, and 7 small dots, and the same numbers of large dots. Each session consisted of four blocks of trials. As already indicated, 2, 3, and 4 dots were positive for two subjects, whereas 5, 6, and 7 dots were positive for the other two birds.

The results are shown as total responses per session to the several patterns representing each numerosity value. The aforementioned discrimination ratio (DR) can be calculated for any two numerosity values, such as 4 and 5, or 2 and 7. The DR obtained with displays of 4 and 5 dots was of particular interest.

Experiment 1:
Discrimination Between Ranges of Small Numbers

The pigeons were trained only on the final discrimination between the two ranges of numbers previously specified. They acquired this discrimination very slowly, and only a selected portion of the data is presented here. The mean total responses obtained from the sixth block of the seven blocks of eight training sessions are shown in Fig. 3.3 for the two birds with the larger numbers positive. The data were quite stable by this time. The pigeons responded differently to the ranges of positive and negative values. Both birds provided a marked gradient within the negative range: They responded more to larger numbers of dots, which were closer to the positive numbers. Differences in responding among the positive values were much smaller. S 5376 did provide a modest gradient among these values with both large and small dots. S 8439 did not respond differentially within the positive range. The two birds differed in their discrimination between 4 and 5 elements. Whereas S 8439 clearly differentiated between the patterns of 4 and 5 elements, S 5376 did not. The sizes of the elements did not exert a large effect among the sets of negative elements. S 5376 responded somewhat less to the smaller numbers of positive elements, which may indicate that this bird was influenced by the total areas of the elements in the array.

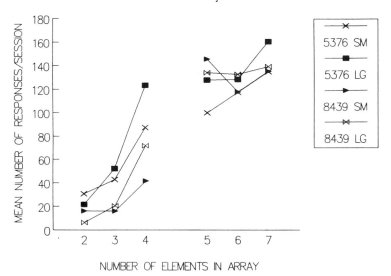

FIG. 3.3. Discrimination between two ranges of small numbers of elements by two birds in Experiment 1, obtained from the sixth block of eight training sessions. The larger numbers were positive. Data obtained from arrays of small (SM) and large (LG) elements are separately presented.

Fig. 3.4 provides corresponding results from two birds with the smaller numbers positive. For both subjects, we see a modest gradient of responding within both the positive and the negative ranges of numbers. But again, there is a marked difference between subjects in the numerosity discrimination that separates the two ranges. S 8468 learned this discrimination very well. However, S 3360 did not respond differentially to the patterns of 4 and 5 elements. Neither subject was affected by the sizes of the elements.

The DRs obtained from the patterns of 4 and 5 elements are shown in Table 3.1, for the seven blocks of eight training sessions. Bird 3360 showed little acquisition of this number discrimination. The performance of 5376 was mediocre at best. The two other birds did somewhat better, but the DRs seldom exceeded .70. (To achieve a ratio of .67, the subject has to respond twice as much to the positive than to the negative arrays.) However, the discriminations were better among the more disparate numerosity values. Birds 5376 and 3360 discriminated poorly between 4 and 5 elements, but they provided orderly gradients of responding within the positive and within the negative ranges of values. Although these subjects discriminated among the numbers of elements, the discriminations did not reflect the contingencies of reinforcement very well. The other two birds (8468 and 8439) did a good deal better in this respect.

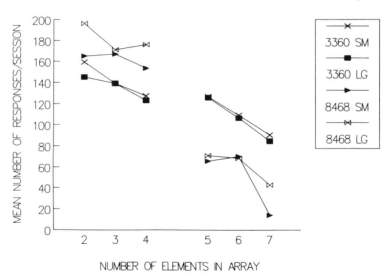

FIG. 3.4. Discrimination between two ranges of small numbers of elements by two birds in Experiment 1, obtained from the sixth block of eight training sessions. The smaller numbers were positive. Data obtained from arrays of small (SM) and large (LG) elements are separately presented.

TABLE 3.1
Discrimination Ratios Obtained from Arrays of 4 and 5 Elements

Pretraining	Condition	Blocks of 8 Sessions							
		1	2	3	4	5	6	7	8
No Pretraining									
3360	2,3,4+					0.50	0.50	(0.50)	
8468	2,3,4+					0.72	0.71	0.72	
8439	5,6,7+					0.67	0.71	0.66	
5376	5,6,7+					0.59	0.52	0.57	
Mean						0.62	0.61	0.61	
		1	2	3	4	5	6	7	8
		0.53	0.52	0.54	0.52				
		0.52	0.60	0.62	0.77				
		0.58	0.65	0.63	0.76				
		0.53	0.55	0.52	0.66				
Mean		0.54	0.58	0.58	0.68				
4 vs. 5		1	2	3	4				
6584	2,3,4+	0.52	0.48	0.55	0.53				
37	2,3,4+	0.56	0.48	0.53	0.53				
8057	5,6,7+	0.45	0.54	0.53	0.51				
5442	5,6,7+	0.56	0.51	0.53	0.51				
Mean		0.52	0.50	0.53	0.52				
						1	2	3	4
						0.51	0.52	0.55	0.51
						0.55	0.64	0.61	0.62
						0.55	0.63	0.55	0.51
						0.55	0.52	0.52	0.65
Mean						0.54	0.58	0.56	0.57
						5	6	7	8
						0.54	0.50	0.49	0.53
						0.65	0.60	0.67	0.68
						0.55	0.55	0.57	0.61
						0.62	0.58	0.64	0.64
Mean						0.59	0.56	0.59	0.61
2 vs. 7		1	2	3	4				
8437	2,3,4+	0.60	0.70	0.69	0.78				
8478	2,3,4+	0.55	0.58	0.64	0.72				
8423	5,6,7+	0.64	0.54	0.56	0.72				
8460	5,6,7+	0.56	0.58	0.65	0.71				
Mean		0.59	0.60	0.64	0.73				
						1	2	3	4
						0.57	0.52	0.63	0.61
						0.53	0.56	0.58	0.54
						0.60	0.64	0.80	0.74
						0.59	0.70	0.66	0.68
Mean						0.57	0.60	0.67	0.64
						5	6	7	8
						0.69	0.62	0.63	0.68
						0.63	0.59	0.54	0.49
						0.73	0.76	0.67	0.63
						0.62	0.67	0.63	0.67
Mean						0.67	0.66	0.62	0.62

Experiment 2: Prior Discrimination of Four and Five Elements

The results from Experiment 1 suggest that the discrimination between adjacent numerosities (the "number discrimination") is difficult, and it may have been facilitated by generalization from the more discriminable values. If so, this would support our conceptualization of numerosity as a dimension of stimulus control. In order to determine whether this generalization did support the number discrimination, we trained pigeons first between the two adjacent numerosity values (4 and 5), and then we extended the range of numbers from 2 to 7. The final procedure, therefore, was identical to the training carried out in Experiment 1.

Method

Four new subjects were trained to discriminate between arrays of 4 and 5 dots. Each number was represented four times by large and four times by small dots in each session. Each session was therefore composed of 16 trials. The birds were trained for 32 sessions on this problem. Not a single subject acquired the discrimination; the data are described in the next subsection. We then extended the training procedure to incorporate the range of numbers used in Experiment 1. The same procedure was used. The birds were trained on the maintained discriminations for 8 blocks of 8 sessions. Although it was of interest to replicate our prior findings with the more discriminable numerosity values (e.g., 2 vs. 7), the more important question was whether such training would enhance the discrimination between 4 and 5.

Results

The data obtained from training with 4 and 5 elements are shown in Table 3.1 for the four blocks of eight sessions before the full range of numerosities was introduced, and for the eight blocks of training after the range of numbers was extended. Clearly, the birds did not acquire the initial discrimination with 4 and 5 elements; the DRs were close to .50, and did not increase in the course of training. When the range of numbers was extended, three of the four birds showed substantial improvement in the discrimination between 4 and 5 elements. Subject 6584 did not benefit noticeably from the change in the training procedure. The other three reached DRs greater than .60 during at least three blocks of sessions, which means that the rate of responding to the positive patterns exceeded the rate of responding to negative patterns by at least 60%.

The three birds that eventually acquired the discrimination between 4 and 5 elements also provided a pattern of stimulus control that rather

3. NUMEROSITY AS A DIMENSION OF STIMULUS CONTROL   73

nicely replicated our prior findings. Data from the sixth block of eight training sessions are shown in Figs. 3.5 and 3.6. The pigeons generally provided declining discrimination gradients within the range of the negative values. The data from pigeon 6584 were less convincing than those from the other subjects. Within the range of positive values, the gradients were less orderly. With the exception of the gradients obtained from S 6584, the data are fairly similar to those obtained when the pigeons were trained initially with the full range of numerosity values (Figs. 3.3 and 3.4).

Discussion

This experiment provided findings of interest for the study of absolute numerosity as a stimulus dimension. For three of the four birds, the addition of the more discriminable values to the training procedure improved the discrimination between the two adjacent values on the dimension. We also replicated in a general way the gradients obtained from small numbers of stimulus elements, including the observation that the gradients are generally flatter within the range of positive values than within the range of negative values. The facilitation of the difficult discrimination by the addition of the more discriminable numbers is related to the "errorless" learning procedure originally devised by Terrace (1963; see also Terrace, 1966). This is an instance of the "easy-to-hard

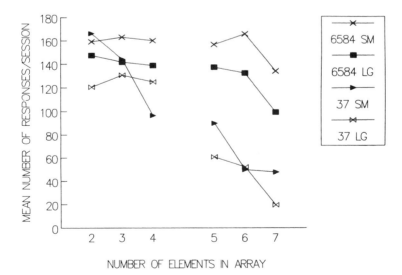

FIG. 3.5. Discrimination between two ranges of small numbers of elements by two birds in Experiment 2, obtained from the sixth block of eight training sessions. The smaller numbers were positive. Data obtained from arrays of small (SM) and large (LG) elements are separately presented.

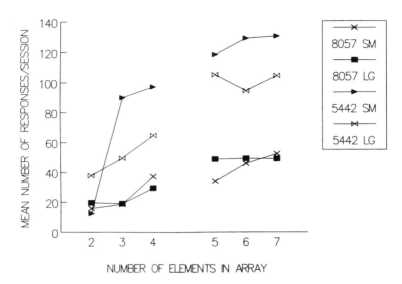

FIG. 3.6. Discrimination between two ranges of small numbers of elements by two birds in Experiment 2, obtained from the sixth block of eight training sessions. The larger numbers were positive. Data obtained from arrays of small (SM) and large (LG) elements are separately presented.

effect." The replication of this effect in the present context supports the conceptualization of absolute numerosity as a stimulus dimension.

Experiment 3: Prior Discrimination of Two and Seven Elements

In Experiment 2, the pigeons discriminated poorly between 4 and 5 elements before the more discriminable numerosity values were added to the training procedure. Although this extension of the range of values facilitated the discrimination between 4 and 5 elements, this may not have been the optimal procedure for this purpose. The birds were trained extensively with that difficult discrimination before the full range of numerosity values was introduced. This initial training may in fact have caused some inattention to the differences between 4 and 5 elements because that discrimination was so difficult.

In the third experiment of this series, we attempted to enhance the number discrimination with a different strategy. The birds were first trained with the most discriminable values, namely 2 and 7 elements. When they acquired this discrimination, the full range of numbers was introduced in the hope that the good discrimination achieved initially would generalize to the other values. This is a version of the "easy-to-hard" procedure.

3. NUMEROSITY AS A DIMENSION OF STIMULUS CONTROL   75

Procedure

Four new pigeons were used for this experiment. Their prior experience as subjects did not involve any discriminations related to the present work. Preliminary training followed the same procedure as Experiment 2, except that arrays of 2 and 7 elements, rather than 4 and 5, were presented. These arrays were shown eight times each in an experimental session, divided equally between patterns of large and small dots. After 32 training sessions, the full range of numerosity values was introduced. This phase was identical to the second phase of Experiments 1 and 2. The birds were trained for eight blocks of eight sessions.

Results and Discussion

Selected discrimination ratios of interest are presented in Table 3.1. The values from the first four blocks of eight sessions are the mean discrimination ratios obtained from initial training with 2 and 7 elements. The four pigeons learned this discrimination reasonably well. During the last block of sessions, the DRs were all above .70. To achieve a DR of .71, responses to the positive stimuli have to outnumber responses to the negative stimuli by a ratio of 2.5 to 1. Clearly this discrimination was easier than the discrimination between 4 and 5 elements obtained in the corresponding initial phase of Experiment 2.

The main data from the second training phase are the DRs obtained from the responses to patterns of 4 and 5 elements, which were introduced for the first time in this phase (together with the patterns of 3 and 6 elements). All subjects showed some initial discrimination between 4 and 5 elements, and this improved to the point where the mean DR was .67 in the third and the fifth blocks. With this ratio, the pigeons were responding twice as much to the positive than to the negative arrays. The discrimination data obtained from the full range of elements are shown in Figs. 3.7 and 3.8 for the eighth block of training sessions.

The pattern of results is rather similar to the discrimination data obtained from Experiments 1 and 2. Three of the birds discriminated well between the patterns of 4 and 5 elements; one (S 8478) did not. The slopes of the discrimination gradients were more marked in the range of the negative than in the range of the positive numerosity values.

DISCRIMINATION AMONG THE NUMEROSITY VALUES: COMPARISONS BETWEEN GROUPS

The number discrimination between 4 and 5 elements is of particular interest, because we expected this might be affected by training with the more discriminable values. All three groups learned this number discrimination to some degree. However, the rates of acquisition, and the levels

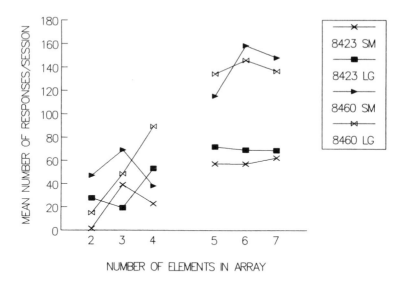

FIG. 3.7. Discrimination between two ranges of small numbers of elements by two birds in Experiment 3, obtained from the sixth block of eight training sessions. The larger numbers were positive. Data obtained from arrays of small (SM) and large (LG) elements are separately presented.

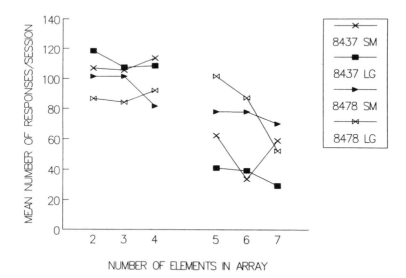

FIG. 3.8. Discrimination between two ranges of small numbers of elements by two birds in Experiment 3, obtained from the sixth block of eight training sessions. The larger numbers were positive. Data obtained from arrays of small (SM) and large (LG) elements are separately presented.

3. NUMEROSITY AS A DIMENSION OF STIMULUS CONTROL

of performance, differed between groups. The mean discrimination ratios from the first seven blocks of training sessions are shown in Fig. 3.9 (and in Table 3.1). On the whole, the group trained initially with 2 and 7 elements learned the number discrimination most rapidly and provided the highest DRs. The other two groups started at a similar and a lower level, but after two blocks of sessions, the control group exceeded the subjects who were pretrained with 4 and 5 elements.

Logarithmic functions have been fitted to these data, as shown in Fig. 3.9. The rate of increase in the discrimination ratio is quite similar for the group that was not pretrained and for the group pretrained with 2 and 7 elements, but it is considerably lower for the group pretrained with 4 and 5 elements. This is evident both from the functions, and from the coefficients of $\log x$ (training sessions). The coefficients are very similar for Experiment 1 (.0984) and Experiment 3 (.0911); these in turn differ substantially from the coefficient for Experiment 2 (.0402).

Admittedly, the obtained values are bumpier than the fitted functions, but it should be remembered that the groups were small, and the means are subject to the vagaries of individual performance. Each group contained one subject that, by the end of training, was still discriminating

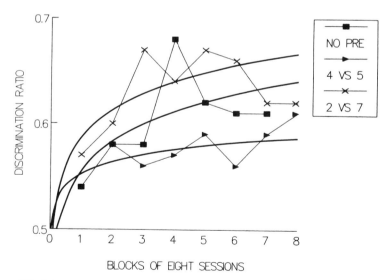

FIG. 3.9. Mean discrimination ratios obtained from blocks of eight training sessions from the three groups of pigeons in Experiments 1, 2, and 3. The subjects were either not pretrained ("no pre"), pretrained with arrays of 4 and 5 elements ("4 vs. 5"), or pretrained with arrays of 2 and 7 elements ("2 vs. 7"). The following logarithmic functions have been fitted to the data from each group: No pretraining: $y = 0.551 + .0984\log x$; 2 vs. 7: $y = 0.584 + .0911\log x$; 4 vs. 5: $y = 0.584 + .0402\log x$.

poorly: 3360 in the no pretraining group, 6584 in the 4 versus 5 group, and 8478 in the 2 versus 7 group.

Extended initial training with the easier numerosity discrimination appeared to facilitate the initial discrimination between 4 and 5 elements, but it did not affect the rate of learning; the fitted functions differ in height, but not in their slopes. Extended initial training with 4 and 5 elements retarded the acquisition of that difficult discrimination, when the more discriminable numerosity values were introduced.

The data presented in Figs. 3.3 through 3.8 for the individual subjects illustrate both the consistencies and the differences in the performance of the small number discrimination by the various subjects. The pigeons varied markedly in their discrimination between 4 and 5 elements. Some learned it well, whereas others did poorly. There is also considerable variability in the degree of differential responding to the numerosity values within the positive and negative sets. Some birds provided clear gradients. For example, Ss 5376 and 8439 from Experiment 1 provided orderly gradients among the negative values, even though one of them (5376) did not discriminate between 4 and 5 elements. On the other hand, S 8437, who discriminated between 4 and 5 elements very well, discriminated poorly within the positive and the negative ranges of numbers.

If the pigeon can acquire the number discrimination between 4 and 5 elements, this suggests that it associates reinforcement and its absence with small numerical values. Therefore, it should also learn that the values within the positive range, and the values within the negative range, were not differentially reinforced. We might then expect that the numerosity gradients would provide "step functions." However, the data from individual birds do not suggest that those subjects that did discriminate well between 4 and 5 elements, also responded at similar rates to 2, 3, and 4 elements, and to 5, 6, and 7 elements. The subjects varied a good deal in their patterns of responding within these ranges, and the differences do not seem to be related to the mastery of the number discrimination.

Given the fact that the pigeons did in many cases respond differentially within the positive and negative ranges, the gradients obtained from 2, 3, and 4 elements should be steeper than those obtained from 5, 6, and 7, because the smaller numbers should be more discriminable. However, the data are quite variable, and do not support this as a consistent finding. The absolute levels of responding differed between subjects, so that comparisons are hard to justify. A more extended analysis would sample different stages of training, and it would be useful to develop an index of the slopes of the gradients.

NUMEROSITY DISCRIMINATION WITH LARGER NUMBERS

The experiments described in this chapter differ considerably from our prior work on the discrimination of relative numerosity (Honig & Stewart, 1989; see also Honig, 1991). In that procedure, all arrays of elements contain the same total number of elements, but the proportions differ, and the pigeons discriminate among those proportions. In the simplest procedure, one uniform array, for example, 25 large dots, is positive, whereas a different uniform array, for example 25 small dots, is negative. (Such contingencies are always counterbalanced.) The pigeons are then tested with different proportions of these elements. Stimulus control has been acquired quite readily on several dimensions with the uniform arrays, and orderly gradients of relative numerosity were obtained with different proportions of elements. The elements have differed in color, form, size, and conceptual category (Honig & Stewart, 1989). The total numbers of elements ranged from 9 to 64 in various studies.

It is perhaps surprising that these discriminations, based on larger numbers of elements, were much easier to acquire than the absolute number discriminations described here with smaller numbers of elements. The difference may lie in the nature of the discrimination: relative numerosity judgments may be easier for the pigeon. On the other hand, the larger numbers of elements may actually have made the arrays more discriminable. With larger numbers, the distances between items are smaller, and may have facilitated the discrimination. In order to clarify this question, we trained pigeons on an absolute numerosity discrimination with arrays consisting of larger numbers of elements.

Method

Four pigeons were trained on a discrimination between 25 and 16 elements. The general training procedures were the same as those previously described, with respect to trial duration, reinforcement, and so forth. The pigeons were trained with the following arrays of 25 red elements: 25 large elements (9.5 mm in diameter on the response screen), 25 medium elements (4 mm in diameter), 25 small elements (3 mm in diameter), and mixtures of 25 large and small elements. The latter were presented in the following proportions: 20 and 5, 15 and 10, 10 and 15, and 5 and 20. The distances center-to-center, between adjacent elements were 13 mm in all arrays.

Training arrays of 16 elements consisted similarly of small, large, and mixed small and large elements. The mixed arrays provided the following proportions: 12 and 4, 8 and 8, and 4 and 12. These arrays of 16 items covered the same area on the response screen as those of 25; however, this resulted in a confound between the number of elements in the array and the distances between adjacent items, which were smaller in the arrays of 25 elements. Therefore, a further set of 16 large elements was prepared in which the distances between adjacent items were the same as those in the arrays of 25. This reduced the overall extent of these arrays of 16 elements.

Each training session consisted of 4 blocks of trials in which one of each of these 8 kinds of arrays was presented. A session therefore consisted of 16 positive and 16 negative trials. The test sessions included 8 additional trials with 9 elements (4 with large dots and 4 with small dots) and 8 additional trials with 36 small elements. Reinforcement was withheld during the test sessions.

Results

The three pigeons that learned the discrimination to a discrimination index of .90 did so rather quickly; each required six blocks of four sessions

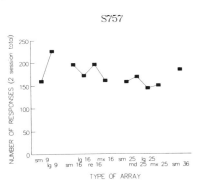

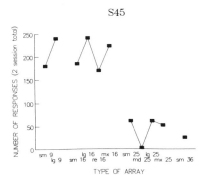

FIG. 3.10. Responses obtained from test sessions with arrays of larger numbers. The pigeons had been trained with 16 elements positive, and 25 elements negative. See the text for details regarding the composition of the arrays. "Sm" means small, "lg" means large, "md" means medium, "re" means respaced.

3. NUMEROSITY AS A DIMENSION OF STIMULUS CONTROL 81

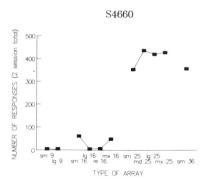

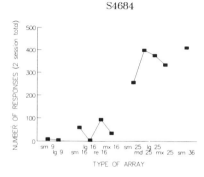

FIG. 3.11. Responses obtained from test sessions with arrays of larger numbers. The pigeons had been trained with 25 elements positive and 16 negative. See the text for details regarding the composition of the arrays. Abbreviations are the same as in FIG. 3.10.

each. The training results were similar for the various ways in which the 25 and 16 elements were displayed, whether they were small, large, or mixed in size. The data from the test sessions are shown for the four subjects in Figs. 3.10 and 3.11. The discrimination between the training values (16 and 25 elements) was well maintained by the three birds that learned it, in the several ways these numbers were presented. The discrimination also transferred well to the test values: from arrays of 16 to 9 elements, and from arrays of 25 to 36 elements. The pigeons were little affected by the internal characteristics of the arrays—the sizes of elements and their spacing. Responding was under the control of the absolute numbers of elements in the arrays.

Discussion

These absolute numerosity discriminations were easier that those we have described at length for small numbers of elements. Large differences in responding to the training values were maintained during testing, and these generalized appropriately to new test values. The absolute difference among the numbers of training elements was of course greater than for the small number discriminations—9, rather than between one (4 vs. 5) and five (2 vs. 7). However, the *proportional* difference between 25 and 16 elements (1.56 to 1) is about the same as between 4 and 6. Perhaps the most interesting comparison involves the initial discrimination between 2 and 7 elements in Experiment 3 of the small numerosity series. The ratio between the numbers of elements is 3.5 to 1. The mean DR, achieved by the pigeons in four blocks of sessions was .73. A DR of .75 reflects a ratio between responses to positive and negative elements of 3 to 1. The three pigeons that learned the discrimination between 16 and 25 elements exceeded that value; the ratios were in the range of 4 or 5 to 1. Other differences between the present experiment and the small number discriminations will be taken up in the General Discussion.

The transfer of the discrimination between 25 and 36 elements, and between 16 and 9 elements, support the conclusion that the discrimination was based on numerosity. However, the confounds based on interelement distances persist with these arrays. Although the data support the present argument for a numerosity discrimination, they do not point unequivocally to this process. Certainly, an interpretation based on a numerosity discrimination is the simplest. And if it is correct, it raises the problem why the numerosity discrimination should be easier with larger than with smaller numbers of elements.

GENERAL DISCUSSION

Absolute numerosity is a derived dimension of stimulus control in the sense that it is an emergent property of sets of *carrier* stimuli. In nature and in the laboratory, these stimuli are presented at the same time or in succession. Differences in numerosity are almost inevitably confounded with other features of the arrays of elements that instantiate them. In a spatial array, for example, the extent of the array increases with the number of elements, if the distance between elements is constant. So does the sum of the *surface area* of the elements. If these confounded features are compensated, this tends to introduce other confounds. For example, if the interelement distances are adjusted to keep the extent of the arrays the same, this introduces a different cue that is correlated with the numbers of items.

One can test the effectiveness of these confounded cues by changing their values while keeping the number of items the same. For example, we could analyze the interitem distance as a distinct variable, rather than randomizing this feature of the arrays, as we did in the present research.

A posthoc analysis would reveal whether interitem distance affects the numerosity discrimination. If it does not, then this would support the notion that this discrimination is actually controlled by the numbers of items. We did control for the total area of items in the area in the main body of the research, by using small and large dots. This seemed not to affect the discrimination in any consistent way. In the experiment with pictures of people, the size of the individual persons was correlated inversely with their number. A posthoc test indicated that the pigeons were not affected by the size of the persons, which supported the conclusion that the discrimination was indeed based on the number of persons.

Computer-generated displays will be particularly useful for the study of various aspects of the displays as they may affect judgments of numerosity. Elements could vary more widely and more randomly in size. Likewise, the distances between elements should be varied in order to study the effectiveness of this cue in arrays that differ in the number of items—whether widely spaced items appear more numerous, because the array is larger, than more closely spaced elements.

Questions of this nature are of systematic importance. Just because a cue is correlated or confounded with the number of elements, this does not imply that animals would be controlled by it. In the real world, this would not be adaptive. If a bird uses a group of three trees as a landmark, the number of trees should not appear to increase when the bird approaches them (as the retinal image of the array increases), or to decrease as bird moves away. If animals do discriminate small numbers of items, irrelevant features of the arrays should not affect the discrimination. The relative locations of items in an array change as the animal moves about, and this should not affect the perception of their numerosity. *Number constancy* would be adaptive. In this respect, some animals may be more reliable than young children. Many introductory psychology texts (e.g. Gleitman, 1991) cite Piaget's findings that numerosity comparisons by young children are affected by the spacing of the items in the array.

On the other hand, other variables that are not necessarily correlated with numerosity may affect the underlying discriminative process without introducing a bias. Interelement distances, for example, may be important. In the three "small numerosity" experiments, we distributed the elements rather widely on the response screen in order to compensate, at least partially, for the confounds between the number of items and the distances between them, and between the number of items and the extent of the array. However, in the real world it may be particularly useful for animals to discriminate numbers of items that are found in clusters, such as eggs in a nest, or berries on a branch. Animals that are gregarious and tend to congregate may be particularly sensitive to the numerosity of clustered items. The numerosity of widely spaced items may be of little importance, in contrast with the usefulness of such items as landmarks. We may have taken an inappropriate approach to the discrimination of small numbers of items, when we spaced the elements rather widely on the response screen in order to control for the total extents of the arrays.

These considerations may be useful in understanding the relative ease with which the pigeons distinguished among the larger numbers of elements—16 versus 25—in comparison to the smaller ones. The larger numbers were rather densely clustered on the response screen. The interitem distance provides a confounded cue with such numbers of elements in the array; however, when the arrays of 16 items were respaced so that the extent of the array was the same as that of 25 items, the discrimination was not reduced. We have not compared absolute numerosity discriminations when small numbers of items are closely spaced and when they are widely spaced; clearly, this should be done to determine whether "clusteredness" is an important feature for this discrimination as well.

These various considerations support the suggestion that absolute numerosity is a valid dimension of stimulus control. Although the values on this dimension are discrete from the experimenter's point of view, the subject may not be attuned to this discreteness. Indeed, much of the present work, such as the number discrimination between 2, 3, 4, and 5, items, points to such a conclusion. We may be able to enhance number and numerosity discriminations in various ways, by making the elements distinctive, by spacing them closely together, by training initially with very different values, and by using a discriminative choice procedure that does not require a reduction in responding.

It is necessary to determine the importance of the uniformity of items in an array for numerosity discriminations. It is possible that the numerosity of differing, distinctive items is easier to discriminate than very similar items. Certainly the individual persons in our first study were not uniform, and this numerosity discrimination (admittedly between very distinct values) was readily acquired. The relative numerosity of items in an array is easily discriminated when these items represent conceptual categories, such as birds and flowers (Honig & Stewart, 1989, Experiment 4). Indeed, if the discrimination of absolute numerosity is a useful process, it may be facilitated with items that represent the real world for the subject.

These various considerations suggest that we need to develop the sensitivity of the research methods designed to examine the process of numerosity and number discrimination. This is nothing new in the area of stimulus control, especially with respect to complex stimuli, patterns of stimuli, or the relations among stimuli. The processes that the animal can bring to bear on the discrimination of numerosity will be more fully revealed only with the development of flexible and appropriate procedures. These will involve different ways of presenting the stimulus arrays, including successive presentation of different numbers of stimuli. It will also be important to use alternative methods of assessing the discrimination, especially those that do not require the absolute reduction of responding to arrays of items that may be attractive to the subject.

ACKNOWLEDGMENTS

Karen E. Stewart provided extensive assistance in the research reported in this chapter, and prepared the figures. She contributed in many ways to this chapter. Patricia Cole and Brad Frankland provided the functions to fit the data from the number discriminations in Experiments 1, 2, and 3. The research was supported by Grant #AO-102 from the Natural Sciences and Engineering Research Council of Canada.

REFERENCES

Capaldi, E. J., & Miller, D. J. (1988). Counting in rats: Its functional significance and the independent cognitive processes that constitute it. *Journal of Experimental Psychology: Animal Behavior Processes, 14*, 3–17.

Church, R. M., & Meck, W. H. (1984). The numerical attribute of stimuli. In H. L. Roitblat, T. G. Bever, & H. S. Terrace (Eds.), *Animal cognition* (pp. 445–464). Hillsdale, NJ: Lawrence Erlbaum Associates.

Davis, H., & Perusse, R. (1988). Numerical competence in animals: Definitional issues, current evidence, and a new research agenda. *Behavioral and Brain Sciences, 11*, 561–615.

Edwards, C. A., & Honig, W. K. (1987). Memorization and "feature selection" in the acquisition of natural concepts in pigeons. *Learning and Motivation, 18*, 235–260.

Gleitman, H. (1991). *Psychology*. New York: Norton.

Herrnstein, R. J. (1984). Objects, categories, and discriminative stimuli. In H. L. Roitblat, T. G. Bever, & H. S. Terrace (Eds.), *Animal cognition* (pp. 233–261). Hillsdale, NJ: Lawrence Erlbaum & Associates.

Herrnstein, R. J., & Loveland, D. H. (1964). Complex visual concept in the pigeon. *Science, 146*, 549–551.

Herrnstein, R. J., Loveland, D. H., & Cable, C. (1976). Natural concepts in pigeons. *Journal of Experimental Psychology: Animal Behavior Processes, 2*, 285–302.

Honig, W. K. (1991a). Discriminability and distinctiveness in complex arrays of simple elements. In M. L. Commons, J. A. Nevin, & M. C. Davison (Eds.), *Quantitative analysis of behavior. Vol. 10: Signal detection* (pp. 103–120). Hillsdale, NJ: Lawrence Erlbaum Associates.

Honig, W. K. (1991b). Discrimination by pigeons of mixture and uniformity in arrays of stimulus elements. *Journal of Experimental Psychology: Animal Behavior Processes, 17*, 68–80.

Honig, W. K., & Stewart, K. E. (1989). Discrimination of relative numerosity by pigeons. *Animal Learning & Behavior, 17*, 134–146.

Lea, S.E.G. (1984). In what sense do pigeons learn concepts? In H. L. Roitblat, T. G. Bever, & H. S. Terrace (Eds.), *Animal cognition* (pp. 263-276). Hillsdale, NJ: Lawrence Erlbaum Associates.

Terrace, H. S. (1963). Discrimination learning with and without "errors." *Journal of the Experimental Analysis of Behavior, 6*, 223-232.

Terrace, H. S. (1966). Stimulus control. In W. K. Honig (Ed.), *Operant behavior* (pp. 271-344). New York: Appleton-Century-Crofts.

Thomas, R. K., Fowlkes, D., & Vickery, J. D. (1980). Conceptual numerousness judgements by squirrel monkeys. *American Journal of Psychology, 93*, 247-257.

Vaughan, W., Jr., & Herrnstein, R. J. (1987). Choosing among natural stimuli. *Journal of the Experimental Analysis of Behavior, 47*, 5-16.

Wasserman, E. A., & Blatt, R. S. (1992). Conceptualization of natural and artificial stimuli by pigeons. In W. K. Honig & J. G. Fetterman (Eds.), *Cognitive aspects of stimulus control* (pp. 203-223). Hillsdale, NJ: Lawrence Erlbaum Associates.

Wesley, F. (1961). The number concept: A phylogenetic review. *Psychological Bulletin, 58*, 420-428.

Wright, A. A., Cook, R. G., Rivera, J. J., Sands, S. F., & Delius, J. D. (1988). Concept learning by pigeons: Matching to sample with trial unique video picture stimuli. *Animal Learning & Behavior, 16*, 436-444.

CHAPTER FOUR

Counting by Chimpanzees and Ordinality Judgments by Macaques in Video-Formatted Tasks

Duane M. Rumbaugh
Language Research Center
Georgia State University
and Yerkes Regional Primate Research Center,
Emory University

David A. Washburn
Georgia State University

INTRODUCTORY CONSIDERATIONS

Whether animals can learn symbols, use them representationally and abstractly, and use them to solve problems remain questions of high interest. Research since the mid-1960s, but particularly during the 1980s, indicates that chimpanzees are capable of mastering and using symbols in ways that parallel human use of words (Savage-Rumbaugh, 1986, 1987; Savage-Rumbaugh, McDonald, Sevcik, Hopkins, & Rubert, 1986; Savage-Rumbaugh, Sevcik, Brakke, Rumbaugh, & Greenfield, 1990). In accordance with the comparative perspective, the bases for human cognitive competence appear to be shared in significant part with our nearest living relatives, the great apes, Pongidae (Rumbaugh, 1990b). Those findings are, in turn, part of a growing number of comparative psychological studies that document impressive cognitive operations of animals and birds in general (e.g., Roitblat, Bever, & Terrace, 1984).

Counting Versus Sensitivity to Quantities

Animals' abilities to respond to different quantities and human abilities for complex mathematics are assumed in this chapter to be part of a continuum of processes provided by evolution of the brain and experience. Research into animals' sensitivities to specific and relative quantities of things and events, presented simultaneously or serially across time, and their abilities to "keep track" of quantities as they accumulate are probably near the lower end of that continuum. Their ability to count, if extant, would be toward, though remain substantially distanced from, the other end of that continuum that provides for mastery of advanced mathematics.

This perspective leaves unanswered, of course, the question as to just what the primitive criteria are for symbolic counting. How will we know that it is operational in animals? At most, its manifestations will be modest, perhaps even reminiscent of trace elements otherwise overshadowed in a complex compound. Notwithstanding, discernment of these manifestations in animals surely will lead to new perspectives regarding brain structure and cognitive function, as well as other perspectives addressing questions of human evolution.

Tactics

Research tactics into the quantity-based skills of primates at the Language Research Center have focused on *productive* counting skills through use of video-formatted tasks. As is discussed in this chapter, these tactics call for the subjects to produce a specified quantity of events as defined by the value of a given trial's *target number*. To date, the things "to be produced" through the subject's use of a joystick-controlled cursor have emphasized (a) the touching of boxes with the consequence that they change in appearance or, in some instances, disappear from the screen, and (b) the placement of cardinal numerals or black dots on each of a number of boxes of varying sizes and presented in random arrays so as to "count out" the quantity called for by the value of the target number for each trial.

The target number's value is subject to change on a random basis across trials and the computerized presentation and automatic recording of the subjects' performances can rule out inadvertent cuing. The procedures require that the subject (a) discern the value of the target number at the beginning of each trial, (b) monitor what has been done since the beginning of each trial, and (c) determine when the "counting" is complete and the trial should be terminated.

The procedures do not allow for correct performance to be based on recognition or labeling of familiar and predictable quantities that recur across trials. Specifically, they do not permit the subject to *label* a familiar visual array of items as "containing *n* items" with a numeral. Rather, they ask the subject to *produce* "a count" on a piece-by-piece or response-by-response basis in response to various target numbers. To the degree that chimpanzees can do so, it is held that true counting skills are being tapped.

Although animals' skills are very circumscribed by comparison to those of humans, this does not detract from their scientific value. On the contrary, that they are extant in *any* measure is of substantial value. As stated earlier, to the degree that such skills in animals can be discerned, we have promise of defining important dimensions of brain-behavior relationships and, also, basic parameters of entry-level numeric competence.

Learning Sets and Caveats

For all who work with the larger-brained monkeys and great apes and for those who are interested in reading of the animals' competencies, a caveat regarding their advanced one-trial, learning-set (Harlow, 1949) skills merits emphasis. It is a well-established fact that great apes and some monkeys (notably the rhesus macaque, *Macaca mulatta*) are capable of learning new discriminations within a single trial *and* also are capable of reversing a learned discrimination within one trial (e.g., Rumbaugh & Pate, 1984). Consequently, the successive readministration of what were novel test trials at the beginning of any test of competence can permit the chimpanzee to relearn and/or modify prior learning so rapidly and efficiently that it can give specious readings of competence, for example with numerals and other symbols. Such performance can be a reflection of discrimination learning-sets rather than ability to use numeric cues with competence. Especially when reward is given for each correct or desired kind of response on so-called test trials, the successive readministration of what were initially novel test trials or tests of generalization affords the perfect opportunity to learn anew via learning sets available to a sophisticated primate.

By definition, novel test trials can be given only once. After their first administration they are, in some measure, familiar and their readministration affords additional training. The use of novel test trials to determine the natural processes underlying primates' performances is highly recommended. Data from those trials should be reported apart from subsequent readministration of the once-novel trials because of the special value they have—and deserve—in the assay of skills. It is performance on the first administration of novel test trials, not on the readministration of them over and over, that provide the purest assay of subjects' competence for the conceptual use of numbers versus their competence for rote associations/pairings between sets/arrangements of items and numbers and vice versa.

Presentations of Testing Materials—Risks

There should be ready agreement to the procedure that *blind* tests should mean the subject cannot be cued by the experimenter's activities in preparing materials for specific trials or by the experimenter's activities while the subject is executing a choice. The risks of cuing are greatly increased when the subject has a small and finite number of things/cards handled and presented personally by the experimenter in an unchanging pattern or order across a span of trials.

Controls and objective methods of inquiry and data capturing have been and must remain the hallmarks of experimental psychology. They should become and remain the hallmarks of all research into the numeric and symbol competencies of both humans and nonhumans. The rhetoric of the research report should never supplant relevant data and the implementation of adequate controls.

Need for Sufficient Amounts of Data

Researchers also should strive to obtain data in amounts sufficient to make tests of significance, in the final analysis, an exercise of the obvious. This effort can tax the creativity of the researcher who, in working with but a few numerals, might soon learn that there is limited option for the construction of novel test trials with which to assay competence after training. Modern technology and particularly the use of computer-based systems can be helpful in solving that and other problems. It is highly adaptable to many of the other needs of research with nonhuman primates (e.g., Rumbaugh, Richardson, Washburn, Savage-Rumbaugh, & Hopkins, 1989; Washburn, Hopkins, & Rumbaugh, 1989).

PERCEPTIONS OF QUANTITIES

Not surprisingly, chimpanzees—hearty eaters that they are—prefer larger rather than smaller portions of foods and drinks (e.g., Menzel, 1960). Survival across eons has perfected the processes whereby primates differentiate relative mass and/or amounts of items almost instantaneously.

How do organisms that have not developed a formal counting system judge relative quantities? What are their limits? How do they do this when the items comprising clusters of quantities are not contiguous? Can they "combine" items of different clusters or groupings so as to judge two or more clusters as containing more than those of others? And if they can do so, are the processes for so doing requisites for addition with numerals? As summarized here, we have investigated these questions, in part, by testing the chimpanzee's ability to choose between pairs of quantities of preferred foods—bits of chocolates.

Summation Studies with Chimpanzees

Two language-trained chimpanzees, Sherman and Austin (*Pan troglodytes;* see Savage-Rumbaugh, 1986, for a review of their language training), gradually came to choose the larger quantity of chocolates when given a choice between two trays, each with a *pair* of food wells containing chocolate chips (Rumbaugh, Savage-Rumbaugh, & Hegel, 1987; see Fig. 4.1). It is important to note they were not required to do so in order to get chocolates. On the contrary, they were always allowed to eat what they chose. Notwithstanding, their choice of the tray with the greater sum remained high (>90%) even when tested with novel arrays and was a function of the ratio formed by the combined number of chocolates of the two trays. The coarser the ratio (e.g., 2:3 vs. 7:8), the more likely it was that the pair with the larger amount would be selected. The findings were interpreted to suggest that the chimpanzees could (a) subitize (e.g., determine specific, small quantities without counting) quantities of the individual wells that formed the right and left pairs, and then (b) sum the paired subitized values, so as (c) to choose the pair that offered the greater total quantity. Summation was defined as the reliable choice of "one of a *pair* of quantities whose overall sum is greater than the sum of another pair of quantities for all possible pairs within a stated numerical range" (Rumbaugh et al., 1987, p. 107).

A detailed assessment was made to determine the degree to which correct performance might be the result of avoiding the pair that had the least number of chocolates in a single well or by selecting the pair that had the greatest amount in a single well, rather than summing across wells (Rumbaugh, Savage-Rumbaugh, & Pate, 1988). These alternative hypotheses were tested by (a) devising pairs such as 4,3 versus 5,3, in which the choice for the greater total could not be made by avoiding the side with smallest single quantity in one of its wells (i.e., each pair had a well with 3 chips) and (b) devising pairs such as 3,5 and 4,5 in which the choice for the greater total could not be made by choosing the pair that had a well with largest quantity (i.e., each pair had a well with 5 chips).

The ratios formed by the summed values per tray ranged from 1:2 to 7:8. Once again, noncorrection testing procedures were employed, meaning that even "error" choices of the smaller sum were rewarded. The chimpanzees *always* got to eat the contents of both wells of whichever pair they selected.

FIG. 4.1. The apparatus used in summation studies.

Choices of the greater sum across conditions ranged from 87%–100% and the distribution of errors for the three conditions were quite similar. No single strategy, other than focusing on the entire array and comparatively summing both sides, could seemingly support choice of the greater sum in the conditions of this study. With these trials randomly interspersed with trials of still other types, it is unlikely that an array of strategies apart from summation could be successfully employed. No signal was given, for example, to inform the subject that on a given trial it was a "select-the-tray-with-the-greatest-single-amount trial" or an "avoid-the-tray-with-the-smallest-single-amount trial," and so on. If the chimpanzees had attempted to use such a group of strategies, they would first have had to evaluate the entire array in some manner to select the strategy appropriate for that trial. This seems quite unlikely, particularly because they always were permitted to eat the chocolates regardless of whether their choice was for the greater quantity. The most parsimonious interpretation inferred that Sherman and Austin used a single strategy across all trials, one based on the summation of quantities per pair as a basis of choice.

More recently, Pérusse and Rumbaugh (1990) replicated the basic phenomena associated with summation and found that summation is manifest even when ratios defined by the two pairs of quantities forms a ratio as fine as 8:9. Through use of varied colors of food bits, it was demonstrated that the choices are not a function of overall brightness associated (or confounded) with the number of chips in each pair of wells. Additionally, the subject does better if it has only two wells, rather than two pairs of wells, between which to choose. This implies that when the chimpanzee chooses between pairs of wells they must negotiate the distance between each pair by some process whereby the quantities of the two wells are pooled, combined, or summed, and this is more difficult than when they have only to negotiate the discontinuity of constituent food bits when they are judging the larger of two clusters.

Skills for summation are viewed as possible precursors to the emergence of formal arithmetic operations, such as addition and subtraction. But, children's entry-level arithmetic operations likely require the ability to count. To pursue research into some possible requisites to formal arithmetic operations, research was designed to explore whether a chimpanzee might be able to count.

CAN A CHIMPANZEE LEARN TO COUNT?

Could a chimpanzee learn to enumerate items and thereby to *produce* specific quantities of responses and attendant consequences in accordance with the values of different Arabic numerals? This question would call for tactics that contrast with those used in studies of *counting* or *number usage* that asked whether an animal could learn one or more associations between Arabic numerals and various arrays constructed and presented by the experimenter. That kind of skill might be reducible to learned associations between numerals and specific arrays that, if correct, nets reward. It would not necessarily entail enumeration and the learning of the values for each of several numerals.

A relevant and important study of earlier years is, of course, the one reported by Ferster (1964), who taught two chimpanzees to match binary digital arrangements with corresponding quantities of items ranging in numbers from 1 to 7. After receiving approximately 500,000 trials, the subjects could reliably label various quantities by producing appropriate binary expressions, regardless of the shape, size, or arrangement of items on the display. Notwithstanding that this was an impressive accomplishment for the chimpanzees, Ferster did not claim that his subjects were counting because there was no indication that they enumerated specific items.

Somewhat similar is a study reported by Matsuzawa (1985) with his language-trained chimpanzee, Ai. Ai used each of 6 keys, glossed as numbers, to label the numeric quantity of arrays comprised of from 1 to 6 items. However, Matsuzawa did not claim that Ai had literally counted the number of items per display. Ai's learning might have been limited to associative couplings between various Arabic numerals and the perceptual patterns defined by various quantities of the items displayed (i.e., toothbrushes, etc.). This consideration does not rule out, however, that Ai might have learned more than that. Indeed, this seems possible when we consider Matsuzawa's report that when the display consisted of objects of two colors, Ai responded with the correct numeral contingent on the color of object "to be counted" as designated, across trials, by the experimenter. (For other recent studies of chimpanzee counting, see Boysen, chap. 2 in this vol.)

Another favorite and important study of note was by Davis and Memmott (1983). Rats inhibited bar pressing for food until three shocks had been received during a 30-minute time period. After the third shock, the rats' bar pressing for food increased markedly, which suggested that the rats were counting the number of shocks—an interpretation that Davis recently moderated. Davis (1984) reported that a raccoon could be taught to select a box containing three items of any sort from an array of cubes containing 1, 2, 3, 4, or 5 items. Davis and Bradford (1986) subsequently found that rats could be taught to travel to a third box regardless of its spatial location. Much earlier, Hicks (1956) reported similar results in the development of a concept of "threeness" in the rhesus *(Macaca mulatta)*. These studies demonstrate animals' abilities to differentiate a single, specific quantity apart from others. Again, however, the distinction between learning to perceive a specific quantity versus counting its constituent items should be kept clear.

Learning to respond to a specific quantity, even accumulated shocks across time, had at one time been thought to reflect a form of counting. According to a simpler and more likely explanation, a recurring experience associated with a context will become established as a "perceptual norm" (Rumbaugh et al., 1987) for that specific context. According to this perspective, whenever a familiar number-based context is encountered, the subject will behave in compliance with the *norm* for that context. The subject, for example, might wait and brace for the final shock of a fixed and predictable quantity (so as to resume eating) or continue to move about (as looking in boxes of varied quantities of things or past boxes in a row) until that norm is *reinstated* or *perceptually matched* (e.g., the requisite shadows/patterns of entrances to tunnels have been perceived). Only then is behavior substantively redirected (e.g., eating is resumed, a choice of boxes is made, an entrance is selected, etc.). But the behavior that transpires as the subject awaits congruence with a context-specific normative experience does not have to entail the counting of events. It might just entail the ability to discern whether a current perception does or does not match the familiar norm based on past experience. (The construct of perceptual norm predates a related one that entails Rosch's view of *prototype* argued by Thomas & Lorden, chap. 6 in this vol.)

By contrast to studies of the type reviewed earlier, the operations of formal counting are quite different. Formal counting is comprised of constituents of interacting cognitive processes. There is no simple criterion for it (criteria for counting discussed later are either from or were suggested by Gelman & Gallistel, 1978). To conclude that an animal counts, one must, among other things, demonstrate (a) *how* it enumerates items of sets, things, or events; (b) that it partitions the counted from the uncounted; (c) that it stops counting appropriately; and (d) that it can count different quantities and kinds of things.

Enumeration has become, quite justifiably, the focal point in studies of counting (Davis & Perusse, 1988), but enumeration must not just be suggested by appearances (i.e., "touching" the items to be counted), which might be ritualistic and without function. Rather, the actual functional role or value of any act of enumeration should be validated. Simply because a child can say "one, two, three" while pointing to objects does not mean the child can count. It must be established that the child can count out just one or two as well as three items and with different kinds and quantities of items and events in various contexts. To be able to enumerate thus implies that competencies for cardinality and ordinality are present.

Cardinality refers to the tags or labels used during counting. If used in the appropriate order, those labels enable counting with precision. By a one-to-one assignment of a tag/label to each item in a set (the one-one principle), one may determine the total count or number of items. But that the subject "knows" that the last item's tag also defines the total items in that set (the cardinal principle) is not necessarily apparent. It must be empirically validated.

Essentially anything might be counted (the abstraction principle). Neither the nature nor quality of items or events determine the basic operations of counting.

Each tag/label used to count has its position or sequence relative to all others. Once counted an item can be said to have been the "third" or the "fifth" counted, and so on. But the *total count* of a set is not influenced by the order or sequence through which the members of that set are counted (the order-irrelevance principle). To count, an organism must know each number's ordinal rank and that each number's cardinal value serves to declare the total quantity counted at each step in the counting process (e.g., the item assigned "three" is the third one and, also, that three items have been enumerated).

Whether a chimpanzee could learn to count, as evidenced by producing the appropriate number of responses to the Arabic numerals 1, 2, and 3, was recently addressed, keeping several of the previous conceptual and procedural caveats at the fore (Rumbaugh, Hopkins, & Washburn, 1989; Rumbaugh, 1990b; Rumbaugh, Hopkins, Washburn, & Savage-Rumbaugh, 1989; Rumbaugh, Savage-Rumbaugh, Hopkins, Washburn, & Runfeldt, 1989). The goal was to have the chimpanzee (Lana, at the time a 17-year-old adult female with an extensive language training history; Rumbaugh, 1977) produce a number of responses in accordance with the values of specific Arabic numerals—1, 2, and 3. She would have to respond so as to produce 1, 2, or 3 discrete units/events—not simply label arrays. In the final test, Lana would have no residual visual feedback or cumulative tally to determine whether she had already produced an array of 1, 2, or 3. Rather, she would have to remember what she had done within the course of each trial in order to complete it successfully. Most training and all test trials would be unique in that (a) the value of the number to which she was to count (i.e., the *target* number), (b) the placement of that number on the monitor, and (c) both the quantity and placement of items available with which she could proceed with the count were randomly determined each trial.

All phases of the present study were run on a Commodore 64 computer with a Sony XY-2400 color monitor. Lana had readily learned how to manipulate a joystick to move a cursor relative to targets on the monitor by observing a researcher. All procedures used to teach Lana the counting behaviors employed her precise control of the joystick and were broken down into three major categories: (a) number orientation and number matching, (b) removal of perceptual or spatial relations, and (c) generalization to new shapes and colors.

The typical trial that Lana saw on the monitor early in her training is shown in Fig. 4.2. Lana's first job was to take the cursor to the target number (which was always 1, 2, or 3). As soon as she did so, the items (selection boxes) with which to proceed with the count then appeared in the bottom half of the screen. Lana's second job was to take the cursor and to attempt to count out 1, 2, or 3 of those items, depending on the value of the target number on a given trial. On the kind of trial por-

4. CHIMPANZEE COUNTING

FIG. 4.2. A trial typical of Lana's early training. The target number is 3 on this trial. Consequently, it appeared in the third box in the row of feedback boxes at the top of the screen. (If the target number had been 2, it would have appeared in the second box from the left, and so on.) As portrayed here, Lana has already removed two of the selection boxes from the lower portion of the screen. When the first one was removed, the number 1 appeared in its stead and the left-most feedback box turned blue; when the second one was removed, the number 2 appeared in its stead and the second feedback box turned blue. The figure shows that Lana now has the cursor on the box in the lower right-hand corner of the screen. A 3 will appear in its stead, and the box containing the target number will turn blue. For the trial to be completed satisfactorily, Lana will next have to return the cursor to that target number.

trayed in Fig. 4.2, as each box was counted a feedback score-keeping box turned blue (beginning with the left-most feedback box in the top row) and a tone sounded briefly. When she set about to terminate a trial, she had to return the cursor to the target number. If she returned to the target number too early or too late, it was treated as an error, in which case she received no reward and the computer sounded a raucous tone.

The number of items in the bottom-half of the screen varied randomly from trial to trial and generally exceeded the value of the target number, so there was no reliable or normative pattern toward which she could work to produce and then use as a basis for terminating the trial. Across the course of 20 software programs, Lana's training always entailed some alteration in the aforementioned display in order to rule out, in final test, the use of all cues and strategies other than counting as the basis for correct performance.

In the final step of the study, Lana's ability to respond to each number without the benefit of residual/cumulative visual feedback (i.e., feedback boxes appearing and/or changing shapes and colors across trials, as in some programs) was tested. Only the tone sounded as each box disappeared when touched by the cursor (Fig. 4.3). Thus, Lana had only her memory of performance within the trial at hand to discern when she should stop the count so as to end the trial successfully.

Figure 4.4. depicts the results from the first 73 no-feedback test trials across which the target numbers were randomly determined. Lana clearly responded significantly above chance for all numbers. Interestingly, she was somewhat better, though not significantly so, on those trials that had no visual feedback than on those that did. It merits emphasis that each test trial was unique in that the position and value of the target number as well as the number and array of selection boxes (from which Lana

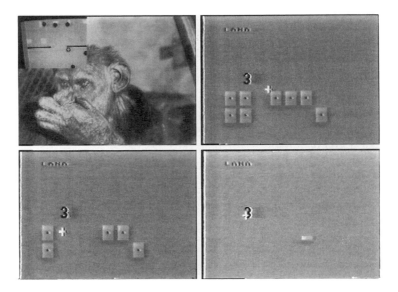

FIG. 4.3. An example of the screen's appearance during a trial on final test of Lana's counting where she had only her memory of intratrial performance to guide her successful performance (e.g., no feedback boxes, as in Fig. 4.2, or other cumulative score-keeping system was present to assist her). The target number might have been 1, 2, or 3. Here, the number is 3 (top right), and Lana must remove any three boxes from those shown in the lower portion of the screen—the quantity and array of which was randomly determined each trial. On this particular trial, she first removed the third box from the left, top row; then the second box from the left, top row; and then the second box from the left bottom row. The array as modified by Lana's performance is portrayed in the lower left of the figure. (A tone sounded briefly when each box disappeared.) She then returned the cursor to the target number for successful completion of that trial (lower right of this figure shows her cursor back on "3" and all but a small portion of one box removed from the screen as the trial ends).

4. CHIMPANZEE COUNTING

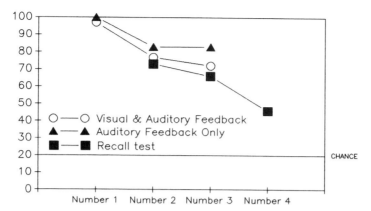

FIG. 4.4. Percentage correct for Lana on visual and nonvisual feedback final test trials for numbers 1, 2, and 3. (See text for further details.) The recall test and the first tests with number 4, randomized with numbers 1, 2, and 3, were given 3 years after the initial study had been conducted.

counted) were randomly determined. Consequently, she could not just remove boxes in order to leave in place a configuration of selection boxes of a given number or of a given array. Since the data of Fig. 4.4 were published, Lana has demonstrated ability to remove 4 boxes in test where the numeral "4" was randomly presented across trials with 1, 2, and 3.

The results of this study are in support of the hypothesis that Lana was counting. Her final test afforded no residual record for her to accumulate as a "score" as she removed, via counting, selection boxes from the lower portion of the monitor. Consequently, she had only her memory of *intratrial* events—what had been done and what had yet to be done on each of series of uniquely configured test trials—as the basis for termination of a given trial. For that to be possible, her skills of enumeration and memory must have entailed elements of both ordination and cardination.

Just how she tagged each box as it was removed to achieve a count equivalent to the value of a given trial's target number is not known. It might have been visual imagery that generated a 1, 2, and 3, in turn, or it might have been something idiosyncratic and quite foreign to representation by Arabic numerals. Notwithstanding, those numerals had been available to serve various functions throughout her training, so it is possible that they served as tags for her entry-level counting skill.

It merits emphasis that Lana was not highly stylized/ritualistic or rhythmic in her removal of selection boxes to achieve the count. From time to time, she would pause midtrial, then resume her work. A review of the 1,442 taped trials from all parts of her 20 training programs revealed that she paused on 303, or 21%, of them. If when she resumed her work she made no change in tactics (e.g., continued the cursor path in progress prior to the pause) she completed 210, or 71%, of the trials correctly—only 10% less than her overall accuracy of 81% on those 1,442 trials. More interesting is the finding that on the 93 trials where Lana changed her tactic after the pause (e.g., changed the direction of the cursor's path, as from a selection box to the target number to finish the trial, or from the target number so as to remove yet one more selection box before completing the trial), she was 61% correct! Consequently, we conclude that whenever Lana altered her tactics *midtrial* (i.e., after pausing), it served on a better-than-chance basis to enhance the probability that she would complete a given trial correctly. This finding suggests that Lana's intratrial memory was one of attentively monitoring events that served as a basis for deciding "what to do next"—to remove another selection box or to return the cursor to the target number.

Some of the occasions for Lana's pausing was due to the activities of other chimpanzees, including the demands and antics of her own offspring, that distracted her within the course of trials. Notwithstanding, she did remarkably well. A related question is: Just how much distraction might Lana tolerate within the course of trial and still be correct? The answer to this would tell us something about the short- and long-term memory operations that supported correct counting.

Her skill with the joystick entailed a very reasonable approximation of a productive enumerative act (i.e., touching the cursor to boxes) that (a) partitioned, in any order, the counted from the uncounted, and (b) served to produce the set/quantity to-be-counted, as declared by the target number's value on each trial. To attempt to reduce Lana's counting to prototypic matching is to deny the construction of her behavior. The "prototype," if one were to insist on positing one, was in Lana's learning of the specific quantity that equaled 1, 2, 3, and 4 boxes *to be removed*. And, that is counting!

We have no evidence to suggest that chimpanzees have a strong "sense of numbers." Although they can discern small differences between quantities, they have not yet given us reason to believe they have any comprehension of the principle that one can always count to a next higher value. The history of ape language research is replete with proclamations that apes cannot do this or that, which subsequently they have been shown able to do, thus we do not say that apes cannot count higher than this or that value.

Although Lana came to count through the course of a series of training programs, not all of them were likely necessary. Quite possibly some phases may have even been too redundant or conservative, and in effect impeded her progress. Notwithstanding, they constituted our first exploration into chimpanzee counting skills with a video-formatted task. They allowed us to move from an initial program that by intent was rich with redundant information to cultivate her counting competence in final test. Gradually, all redundant, all nonnumeric cues, and all residual and obvious accumulative methods whereby Lana might have been "keeping score within trials" were deleted. What remained was a competence to respond to each of three Arabic numbers in a differential and relatively accurate manner—in a manner that we term *entry-level counting*.

It is readily acknowledged that the counting skills of humans clearly entail many more operations than those acquired or demonstrated by Lana. Research with her counting skills are limited in that they have yet to be tested in other contexts. Whether they can provide for the labeling of various sets appropriately has not been tested.

Research in progress with Lana and other language-experienced chimpanzees (Sherman and Austin) supports the view that they can count at least up to 4 or 5 boxes by placing either Arabic numerals or black dots on boxes in accordance with the value of a given trial's target number (Fig. 4.5; Rumbaugh, 1990a, June).

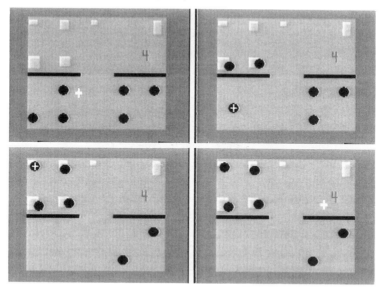

FIG. 4.5. In this trial of the dot task with a target numeral of 4, Lana must place a dot on each of four boxes (upper left and right, then lower left of figure) before terminating the trial by returning the cursor to the target numeral (lower right).

RHESUS MONKEY COMPETENCIES FOR ORDINALITY

Washburn and Rumbaugh (1991) reported that through the use of video-formatted training that rhesus *(Macaca mulatta)* learned the relative values of Arabic numerals 0 to 9. The ability of rhesus to use the joystick with great skill in video-formatted tasks, on the Language Research Center's Computerized Test System (LRC-CTS), provided a ready means whereby they could be presented with all paired combinations of numerals within selected ranges and be permitted to choose one on each trial. As a consequence of selection of a numeral, a corresponding number of pellets was automatically dispensed for the subject to eat. For example, if on a given trial the rhesus had a choice between the numerals 3 and 5 and selected the 3, it received 3 pellets—the delivery of each one being accompanied by a brief tone. If, on the other hand, it selected the numeral 5, it received 5 pellets.

Initial training was limited to the numerals 0 through 5, with a 0 being one of the paired numerals every other trial. If 0 was selected on a given trial, no pellets were delivered to the subject. Within only 900 trials the animals were well above chance.

Subsequent training included numerals 6 through 9. Both rhesus monkey subjects reliably came to prefer the larger numeral of each pair—even 9 over 8. On final test with the administration of novel pairs withheld from prior training (6,5; 7,6; 9,8; 6,4; 7,5; 9,7; and 8,5), one monkey was without error on their first presentation, hence significantly better than chance.

The temporal intervals between the delivery of successive pellets were then systematically varied and were concluded not to have been a determinant of choice of numerals. (Intervals were such that the delivery of 2 or 3 pellets might span a longer time than the delivery of a greater amount, such as 8 or 9.)

Thus, even rhesus monkeys have the well-honed ability to differentiate quantities of things that occur even serially across time. They also can learn quite readily to associate arbitrary symbols (i.e., Arabic numerals) with different quantities and select symbols preferentially on the basis of their relative food-values. Their mastery of the numerals' ordinal ranks was so complete that it readily transferred to an array of numerals simultaneously presented (e.g., when the monkeys had five randomly selected numerals presented simultaneously in random arrays across trials, they "counted down" by selecting with high reliability the highest numeral remaining on the screen—though they were not required to do so to get rewarded). In response to these arrays both animals significantly chose the numerals in a descending sequence—9, 8, 7, 6, and so on.

4. CHIMPANZEE COUNTING   103

Were the rhesus counting? We do not argue that they did because there was no evidence the monkeys were enumerating pellets. Neither have we discerned that they had learned the cardinal values of the numerals and that each numeral of the sequence differed only by a value of 1 pellet. (Lana had learned that a target numeral of 2 required "one more" than 1 and the a target numeral of 3 required "one more" than 2.) Notwithstanding, the monkeys chose the numerals in reliable accord with their ordinal ranks, which is a necessary, though not sufficient, condition for it to be concluded that counting was operative.

Exactly how the monkeys were able to respond even to novel combinations of arabic symbols in accordance with the numerals' ordinal values remains a question for ongoing research. Performance on novel pairs eliminates the possibility that the monkeys had simply amassed 45 discrete-pair associations—selecting the 7 in a 7:6 pair, for example, because "7 is the correct answer." However, the judgments may have arisen from a matrix of knowledge about relative values, for example "7 is greater than 6, and 7 is greater still than 5" and so forth. It is this matrix of relative values—the knowledge that any two numerals differ by a specific degree (number or amount) and the ability to respond reliably to numerals based on both the difference and the degree of difference—that is the main support of the claim for ordinal judgments by the monkeys.

Alternatively, of course, the monkeys may have perceived the numerals as representing absolute quantities or hedonic values (see Thomas, this volume; e.g., a larger quantity of pellets is more pleasurable/gratifying than is a smaller quantity) rather than as members of an ordinal series, just as they might prefer to select red over yellow over green, and so forth. Indeed, we have reported similar preferences for task selection (Washburn, Hopkins, and Rumbaugh, 1991) that do not imply an ordinal sequence of tasks. However, when these preferences map perfectly and reliably on the critical dimension that distinguishes the members (e.g., wavelength for color, difficulty for task, or number of pellets as in our paper), one is strongly inclined to cite this dimension as the basis for judgment.

Granting then that the monkeys based their responses on number (or amount, as we noted), we do not wish to discount at all the possibility that the judgments may have reflected knowledge of absolute number, or that indeed this may even appear to be the more parsimonious interpretation. However, judgments of theoretical parsimony can be tricky, particularly when Morgan's canon meets Occam's razor: Although knowledge of absolute number introduces fewer conceptual steps or operations, the operations themselves are more complex than those required for mastery of a matrix of relative-order associations. Indeed, the cognitive competencies of the monkeys would seem to be most flattered if responses are interpreted as based on conceptual comparison of absolute quantities ("9 > 8") rather than on acquired knowledge about order ("9 comes after 8").

We have based our claim for ordinal judgments on the monkeys' ability to assert, for example, that 9 > 8 > 7 > 6 > 5. In point of fact, the monkeys generally judged, for example, that 9 > 8, 7, 6, and 5; that 8 > 7, 6, and 5; that 7 > 6 and 5; and that 6 > 5. We see neither the logic nor the parsimony of distinguishing between these two assertions, nor between them and "9 comes after 8 which comes after 7 . . .", although other investigators may wish to maintain the distinctions. Thus, we risk that our judgment of relative parsimony may differ from that of other investigators—confident that in either case the findings are noteworthy. It is clear that the monkeys made precise relative numerousness judgments across a range of ordinal symbols differentiated by a single pellet, and that, regardless of whether the monkeys made judgments based on knowledge of absolute number or relative order, responding was not limited to any simple learning of the "correct" choice for each possible array of numerals.

A PERSPECTIVE

Studies summarized in this chapter serve to emphasize nonhuman primates' abilities to learn symbols' values and to use them to adaptive advantages. Their skills in this regard have been underestimated, in measure because of past technological limitations placed on researchers. With the advent of computer technology, new and efficient research tactics, that provide for excellent controls and replication of studies between laboratories, promise to yield important and more accurate assays of nonhuman primates' abilities for symbols and numbers and more accurate definitions of what it means to count.

ACKNOWLEDGMENTS

Preparation of this chapter was supported by grants NICHD-06016 and ARB-00165 from the National Institutes of Health by grant NAG2-438 from the National Aeronautics and Space Administration to Georgia State University, and by support from the College of Arts and Sciences, Georgia State University. Appreciation is expressed to Terry Fry, William D. Hopkins, and Robin Peters for their assistance and colleagueship.

REFERENCES

Davis, H. (1984). Discrimination of the number three by a raccoon *(Procyon lotor)*. *Animal Learning & Behavior, 12,* 409-413.
Davis, H., & Bradford, S. A. (1986). Counting behavior by rats in a simulated natural environment. *Ethology, 73,* 265-280.
Davis, H., & Memmott, J. (1983). Autocontingencies: Rats count to three to predict safety from shock. *Animal Learning & Behavior, 10,* 95-100.
Davis, H., & Perusse, R. (1988). Numerical competence in animals: Definitional issues, current evidence, and a new research agenda. *Brain and Behavioral Sciences, 11,* 561-579.
Ferster, C. B. (1964). Arithmetic behavior in chimpanzees. *Scientific American, 210,* 98-106.
Gelman, R., & Gallistel, C. R. (1978). *The child's understanding of number.* Cambridge, MA: Harvard University Press.
Harlow, H. F. (1949). The formation of learning sets. *Psychological Review, 56,* 51-65.
Hicks, L. H. (1956). An analysis of number-concept formation in the rhesus monkey. *Journal of Comparative and Physiological Psychology, 49,* 212-218.
Matsuzawa, T. (1985). Use of numbers by a chimpanzee. *Nature, 315,* 57-59.
Menzel, E. (1960). Selection of food by size in the chimpanzee and comparison with human judgments. *Science, 131,* 1527-1528.
Pérusse, R., & Rumbaugh, D. M. (1990). Summation in chimpanzees *(Pan troglodytes):* Effects of amounts, number of wells and finer ratios. *International Journal of Primatology, 11*(5), 425-437.
Roitblat, H. L., Bever, T. G., & Terrace, H. S. (1984). *Animal cognition.* Hillsdale, NJ: Lawrence Erlbaum Associates.
Rumbaugh, D. M. (Ed.). (1977). *Language learning by a chimpanzee: The LANA project.* New York: Academic Press.
Rumbaugh, D. M. (1990a, June). *Chimpanzees: Language, speech, counting, and video tasks.* Paper presented at the American Psychological Society's Convention, Dallas.
Rumbaugh, D. M. (1990b). Comparative psychology and the great apes: Their competence in learning, language, and numbers. *The Psychological Record, 40,* 15-39.
Rumbaugh, D. M., Hopkins, W. D., & Washburn, D. A. (1989, November). *Judgments of relative numerosity by macaques (Macaca mulatta) in a video-task paradigm.* Paper presented at the 19th Annual Psychonomic Society Meeting, Atlanta.
Rumbaugh, D. M., Hopkins, W. D., Washburn, D. A., & Savage-Rumbaugh, E. S. (1989). Lana chimpanzee learns to count by "Numath": A summary of a video-taped experimental report. *Psychological Record, 39*(4), 459-470.
Rumbaugh, D. M., & Pate, J. (1984). The evolution of primate cognition: A comparative perspective. In H. L. Roitblat, T. G. Bever, & H. S. Terrace (Eds.), *Animal cognition* (pp. 569-587). Hillsdale, NJ: Lawrence Erlbaum Associates.
Rumbaugh, D. M., Richardson, W. K., Washburn, D. A., Savage-Rumbaugh, E. S., & Hopkins, W. D. (1989). Rhesus monkeys *(Macaca mulatta),* video tasks, and implications for stimulus-response spatial contiguity. *Journal of Comparative Psychology, 103,* 32-38.
Rumbaugh, D. M., Savage-Rumbaugh, E. S., & Hegel, M. (1987). Summation in the chimpanzee *(Pan troglodyte). Journal of Experimental Psychology: Animal Behavior Processes, 13,* 107-115.
Rumbaugh, D. M., Savage-Rumbaugh, E. S., Hopkins, W. D., Washburn, D. A., & Runfeldt, S. A. (1989). *Lana chimpanzee (Pan troglodytes) counts by Numath* [Videotape]. University Park, PA: Pennsylvania State University.
Rumbaugh, D. M., Savage-Rumbaugh, E. S., & Pate, J. (1988). Addendum to "Summation in the Chimpanzee *(Pan troglodytes)." Journal of Experimental Psychology: Animal Behavior Processes, 14,* 118-120.

Savage-Rumbaugh, E. S. (1986). *Ape language: From conditioned response to symbol.* New York: Columbia University Press.

Savage-Rumbaugh, E. S. (1987). A new look at ape language: Comprehension of vocal speech and syntax. In Richard A. Dienstbier & Daniel W. Leger (Eds.), *Nebraska symposium* (Vol. 35, pp. 201–244). Lincoln, NE: University of Nebraska Press.

Savage-Rumbaugh, E. S., McDonald, K., Sevcik, R. A., Hopkins, W. D., & Rubert, E. (1986). Spontaneous symbol acquisition and use by pygmy chimpanzees *(Pan paniscus)*. *Journal of Experimental Psychology: General, 115,* 211–235.

Savage-Rumbaugh, E. S., Sevcik, R. A., Brakke, K. E., Rumbaugh, D. M., & Greenfield, P. (1990). Symbols: Their communicative use, comprehension, and comination by bonobos *(Pan paniscus)*. In C. Rovee-Collier & L. Lipsitt (Eds.), *Advances in infancy research* (pp. 221–278). Norwood, NJ: Ablex.

Washburn, D. A., Hopkins, W. D., & Rumbaugh, D. M. (1989). Automation of learning-set testing: The video-task paradigm. *Behavioral Research Methods, Instruments, & Computers, 21*(2), 281–284.

Washburn, D. A., & Rumbaugh, D. M. (1991). Ordinal judgments of numerical symbols by macaques *(Macaca mulatta)*. *Psychological Science, 2*(3), 190–193.

PART TWO

COUNTING: CRITERIA AND RELATIONS TO BASIC PROCESSES

CHAPTER FIVE

Numerical Competence in Animals: Life Beyond Clever Hans

Hank Davis
University of Guelph, Ontario, Canada

Numerical competence in animals has many faces. There are considerable differences, not only in the methodology we use to study it, but also in the processes different test situations are likely to invoke.

This simple position has a well-entrenched adversary: If one employs numerical stimuli, the animal must be counting. My own research and reviews of other people's research suggest otherwise. There seem to be a variety of processes that animals can use when dealing with numerical stimuli. Counting is just one of them, and perhaps the least likely.

RELATIVE AND ABSOLUTE NUMEROSITY

I will return to the question of why I believe counting is unlikely, and why Pérusse and I proposed the category *proto-counting* as a way to describe suggestive but incomplete evidence of counting (Davis & Pérusse, 1988). For now, let me make an important distinction between absolute and relative numerical abilities. I believe that *absolute numerosity* is a distinctly human invention. No nonhuman animal needs this form of numerical competence in order to lead a successful, totally normal life. A rat can be a model of his community, as can a raccoon, an African Grey parrot, even a chimpanzee, and never show a trace of absolute numerical competence.

On the other hand, there are what may be broadly classed as *relative numerosity* skills. Perhaps the simplest is the distinction between presence versus absence. Teaching a chimpanzee to discriminate between some grapes and none is hardly impressive and holds a tenuous link to numerical competence, as we generally think of it. Yet, like binary code, presence versus absence may be viewed as the simplest possible building block with which information can be processed or a sophisticated number system may be constructed.

The next level of relative numerosity skills involves the discrimination between or ordering of two or more quantities. A simple example is a more versus less *relative numerousness judgment* (RNJ). RNJs are typically classed as numerical (e.g., Davis & Memmott, 1982; Davis & Pérusse, 1988), although their relationship to absolute numerical ability is questionable (see Davis, Albert, & Barron, 1985). Far more importantly, RNJs may be fundamental to the life of many animals because they are at the cornerstone of foraging behavior. Whether RNJs represent something cognitive or result from simple perceptual processes remains to be determined.

THE NATURALNESS OF NUMBER

Saying that animals are not natural mathematicians may seem an odd statement from someone who has spent the past 10 years investigating numerical competence in animals (e.g., Davis, 1984, 1989; Davis & Albert, 1986; Davis & Bradford, 1986, 1987; Davis, MacKenzie, & Morrison, 1989; Davis & Memmott, 1982, 1983). It is for this reason that I stress that my statement does not mean that animals are incapable of demonstrating many forms of numerical competence under supportive conditions. Rather, it is my belief that such supportive conditions are necessary for successful demonstrations with nonhuman subjects.

What does it mean to say that absolute number has no natural utility for nonhuman animals? Cynically speaking, it means that once again human investigators have taken some characteristic human ability (such as language) and gone looking for rudiments of it in other species. Certainly, there is an arrogance or anthropocentrism to this activity. But just as certainly, this search remains a mandate of comparative psychology.

Many of the contributors to this book have demonstrated numerical competence in their subjects. By implication, this means they have, one way or another, devised ideal environments for exploring numerical competence. Such situations may entail optimal or natural test conditions or optimal rapport between the trainer and subject: Either may be capable of motivating an animal to show evidence of numerical competence.

In short, I do not believe that demonstrations of numerical competence come easily. They are no mean feat, and for each of the successes you have heard about, there are untold failures. Although the nondissemination of negative evidence is the way science normally progresses, it is particularly unfortunate in the case of numerical competence in animals because it clouds the question of how general or easily established this ability truly is.

NUMERICAL COMPETENCE VERSUS COUNTING

To say that we have found evidence of numerical abilities in nonhumans does not imply that we have observed cognitive or mental processes equivalent to those used by humans. I have frequently argued to avoid using the word *counting* in those situations where it is applied merely as a general synonym for numerical competence, with no attempt to specify the underlying process. The uncritical use of the word *counting* seems to be on the wane, although some investigators continue to invoke it "if the number of events is the discriminative stimulus" (e.g., Church & Meck, 1984, p. 448).

Let me offer an analogy to the question of whether animals can count. Griffin (1984) raised the question, "Do Animals Think?", and offered an entire book's worth of what he considered affirmative evidence. Many animal behaviorists, cognitive ethologists, and researchers in animal learning do not share Griffin's conclusion. That in itself is not a problem. What is problematic, perhaps even perilous to his case, is Griffin's dependence on *affirming the consequent* to structure his case for "animal thought." Most of Griffin's evidence takes the form, "If animals were thinking, they would do the following. Animals *are* doing the following; therefore, they are thinking."

In addition to using specious logic, Griffin's case is also flawed by his lack of a rigorous or formal definition of "thinking." The absence of such a central definition not only plagues his book, but was underscored during an address to the Animal Behavior Society (Griffin, 1987). When asked whether he was concerned about the lack of clear criteria for animal thought, Griffin replied, in essence, "Let's stop worrying about definitions of thinking and get on with the business of looking for it."

How precisely does this scenario fit "animal counting"? Certainly, most of us hope the approximation is not at all good. I would like to agree, but there are some troublesome signs. Like all areas of animal cognition, those studying numerical competence must be especially vigilant. We not only have the strictures of behaviorism to contend with, we also have a very unique history. Clever Hans, the putative counting horse, has become part of psychology's racial memory and tainted the field of numerical competence almost to the point of nonrecovery. We have the German scientist Otto Koehler (1950) to thank for helping to wean questions about numerical abilities in animals back into the realm of respectable research. Many procedures and species have been used during the past half century of numerical competence research (see reviews by Davis & Memmott, 1982; Davis & Pérusse, 1988; Rilling, chap. 1, this vol.; Salman, 1943; Wesley, 1961). Although we may admire the diversity and ingenuity of these demonstrations, we must nevertheless ask two basic questions whenever claims for numerical control have been made:

1. Is it the *numerical* feature of the stimuli that is responsible for the behavior in question? This may seem an obvious question, but there are many other features of the environment that covary with number. I believe a strong case has been made for the proposition that such other features (e.g., time, density, size) may be easier, more salient, or more natural for an animal to process. This view is not uncontested (e.g., Capaldi & Miller, 1988b) and is something we must continue to explore. For now, we must be sure that every possible control group has been run. No matter how "natural" we believe number may be for an animal, we cannot ignore nonnumerical alternates. I will return to this argument.

2. Even when nonnumerical features of the situation have been controlled, we cannot assume that the numerical process underlying such control is counting. Counting is not a simple process. It is defined very precisely in the human literature and entails a number of steps that may not come easily to nonhumans. Counting implies, at the least, that the counter have available an ordered series of tags that can be applied in 1:1 correspondence to items one wishes to count. Such tags need not be linguistic in nature, but they do need to occur in stable ordered form. Both Davis and Pérusse (1988) and Gelman and Gallistel (1978) allowed for the possibility of unconventional and nonlinguistic tags and termed them *numerons*. In an earlier analysis of the same situation, Koehler (1950) discussed the possibility that animals were "thinking unnamed numbers" while they processed quantity.

One of the more persuasive themes in the critical commentary that accompanied the Davis and Pérusse review of numerical competence stressed the possibility that animals may use some alternate form of numerical process, unknown or unavailable to humans. The criticism was made that if we treat human counting as the universal standard of numerical competence we may (a) underestimate the abilities of other species, just as we have underestimated the abilities of infant humans in the past, and (b) fail to detect evidence of an alternative process because it does not follow the rules of human counting. In sum, the arrogance implicit in comparative psychology's relentless search for the rudiments of human abilities may cost us an insight into another species' cognitions simply because it has no counterpart in human experience (see commentaries by Johnson, 1988; McGonigle, 1988; Pepperberg, 1988).

ALTERNATIVE EXPLANATIONS

Every scientist who repeatedly studies some form of animal cognition eventually evolves a strong belief or bias about that behavior. Inevitably, strongly held opinions hold the seeds for conflict, especially in an emerging discipline (see related discussion by Rilling, chap. 1, this vol.). Such a bias may take the form that rats have remarkable spatial abilities and use them widely, or that language lies within the capabilities of nonhuman primates. The latter case, for example, would generate disagreement with a colleague who believed that linguistic abilities are merely artifacts of faulty procedures or incomplete data analysis.

The formation of a strong belief need not be detrimental to good science. Ideally, biases emerge as evidence accumulates, shift as evidence itself shifts direction, and continue to shape future investigations. Strong conclusions present problems only when they constrain the outcomes we are willing or able to consider. Such "myopia" might dissuade an investigator from seeing alternatives to his bias-confirming account of the data.

When scientists publicize a conclusion about some form of animal behavior, it is customary to have that view discussed and appraised by their peers. This scrutiny is part of the checks and balances inherent in the scientific method, whether the subject matter is "cold fusion" or numerical abilities of animals. Occasionally, however, an investigator may be actively opposed to the suggestion of alternative explanations, per se. This position goes to the heart of how scientific investigations proceed and there is an example of it within the numerical competence literature.

In a systematic body of experiments, Capaldi and his colleagues concluded that rats could use numerical cues to anticipate the delivery of reward or nonreward in a series of runway trials (e.g., Capaldi & Miller, 1988a; Capaldi & Verry, 1981). The data from these experiments supported the senior author's well-publicized view that "rats and . . . higher animals in general are highly disposed to count reinforcing events and do so routinely, perhaps automatically . . . and as a first resort" (Capaldi & Miller, 1988b, pp. 15–16).

In a subsequent review, Davis and Pérusse (1988) acknowledged this conclusion, but observed that the serial pattern approach of Capaldi and his colleagues would present far more persuasive evidence of counting if every effort were made to preclude rhythmic regularities in event occurrence, ideally through the use of variable and relatively long inter-event intervals. (p. 576)

The reasoning behind this criticism was simple: (a) We did not share Capaldi's general view about the ease with which rats count or use numerical information; and (b) we believed that the procedures used by Capaldi and his colleagues allowed the possibility that subjects might detect and use cues related to the regular temporal pattern with which food was presented.

Our proposal of rhythm as an alternative explanation seemed reasonable. To begin with, other investigators have identified precisely the same variable as a potential confound in studies of numerical competence (see Brackbill & Fitzgerald, 1972; Rilling, chap. 1, this vol.). Furthermore, Davis, Memmott, & Hurwitz (1975) already provided an account of how the discrimination of temporal regularities could exert profound behavioral control in a variety of situations.

Capaldi rejected both the specifics of our account (Capaldi & Miller, 1988b), and also disputed the appropriateness of our having proposed an alternative to his view (Capaldi, 1988a). The former is a legitimate part of the dialectic of scientific inquiry. In fact, Capaldi (1988a) reported that he has since collected data showing that rhythm is not a viable account for his results. We are pleased that such control data will soon be made archival because they strengthen Capaldi's claims about numerical competence, as well as butress the general perception of the rigor of numerical competence research.

However, we are far less sanguine about Capaldi's strenuous objection to alternatives being proposed to his accounts. Indeed, Capaldi (1988a) argued that "the major impediment to taking the number variable seriously today is looking for alternative explanations in counting experiments." In his view, "Control groups are free for the asking. There is an infinite number of them out there." Capaldi argued that capriciously suggested control groups may begin to plague the field. He noted that "alternative explanations may not be based on any particular evidence. For example, we are not even sure the animal can utilize the cue that is being suggested as an alternative." Capaldi (1988) particularly opposed this form of science because "the individual makes his criticism, does not test the effects of this variable to even see if they are likely, and then moves on to something else." Thus, Capaldi (1988a) proposed that "those who suggest control groups might consider running the experiment themselves." Ultimately, Capaldi's concern is that "if everyone who sees . . . (a missing control group) . . . is free to suggest it, we're going to undermine the field."

5. LIFE BEYOND CLEVER HANS   **115**

This is a very strong position and, insofar as it speaks to how the area of numerical competence will evolve, I want to formally oppose it. In my opinion, skepticism is at the bedrock of science. I see no virtue in restricting criticism or confining it to those outside the area. Although criticism from within may threaten the solidarity needed to draw the area into the mainstream, there is a case to be made that diversity of opinion about numerical competence in animals will bring *hybrid vigor* to the area. In my opinion, if we all shared the view that "animals count routinely and automatically," the field would be weakened, not strengthened. Granted, the practitioners of "animal counting" could tacitly agree to support each other's work uncritically until the area is better established; however, such a conspiratorial pact would not only be anathema to good science, but it could ironically lead to the very external criticism that its practitioners were seeking to avoid.

Consider the credibility of research in extrasensory perception (ESP). This is an area in which strongly held and uniform opinion predominates. Not only is external credibility low, but much ESP research suffers from the lack of adequate control groups for alternative mechanisms. The "fortress mentality" of ESP researchers makes them particularly vulnerable to an external critic like James ("The Amazing") Randi, whose public refutations of the area have been devastating (Jaroff, 1988).

There is a suggestion in Capaldi's argument that we are in danger of opening a floodgate of capricious control groups. I agree that such an outcome would undermine any body of research. But what suggestion is there that a wave of arbitrary criticism is likely to emerge? In large measure, this concern stems from Capaldi's (1988a) conclusion that "the number variable has been historically overlooked by learning theorists such as Hull and Crespi." However, to date, only two substantive criticisms seem to have occurred in the modern numerical competence literature. One, ironically, is by Capaldi himself, who proposed a thoughtful addition to research by Rumbaugh, Savage-Rumbaugh, and Hegel (1987) (see Rumbaugh, Savage-Rumbaugh, & Pate, 1988). The other example is the previously cited "rhythm" suggestion that Capaldi tested and dismissed.

I believe that those of us studying numerical competence are often best able to propose credible alternatives to each other's explanations. As for the question of who should run missing control groups, I believe there is no one more appropriate to do so than the original investigator. Alternative accounts can best be tested by introducing a single new independent variable. In all likelihood, replicating a procedure in a new laboratory, with a different experimenter, different strain of animals, and different apparatus will introduce a host of potential variables whose effects may obscure those of a single control condition.

BIASES AND CONTROL GROUPS

The identification of missing control groups is not based solely on deductive logic. There is truth in Capaldi's (1988a) claim that "there is an infinite number (of potential control groups) out there." Science in general, not just the study of numerical competence, would disintegrate into triviality if logic alone dictated the necessity of control groups. On a strictly logical basis, for example, one might criticize my work by noting that I have only demonstrated numerical competence in Canadian rats or in the months of October and November. Obviously, the suggestion of which control groups to run must be tempered by judgment. It is here that our strongly held opinions help us select among a range of variables whose effects need to be controlled. Just as biases may keep an investigator from seeing the need for a control group, so may strongly held positions prompt an investigator to suggest an alternative account or discount others. Capaldi (1988a) underscored this possibility, noting the biased historical assumption that "whenever a number event has been confounded with a non-number cue, the non-number cue will overshadow it."

This is a fair indictment of the conclusion that has resulted from much of my own research and no doubt influenced our suggesting the need to control for rhythm. My own bias is the result of extensive research showing that, in general, whenever a simple exteroceptive stimulus or traditional stimulus contingency (e.g., *if tone-then food*) was available, it overshadowed or took precedence over more subtle cues or stimulus relations, such as those involving numerical relationships. Our theoretical treatment of *autocontingencies* (Davis, Memmott, & Hurwitz, 1975) surveyed this pattern in our own research as well as that of other investigators. In our earliest experiments, we found that reliable temporal cues used to predict the occurrence of shock or nonoccurrence of food were consistently overshadowed by external predictors of those events (Davis, Herrmann, & Shattuck, 1979; Davis & McFadden, 1978; Davis, Shattuck, & Wright, 1981). Later experiments examined how the same pattern occurred when numerical predictors were stacked against the use of, for example, warning tones (Davis & Memmott, 1983). Although numerical cues were capable of exerting control over behavior, they did so only when more salient predictors (including temporal cues) were precluded. Although we have rarely failed to demonstrate numerical control (see Davis & Albert, 1987 for an exception), our efforts have usually required some form of *environmental sterilization* to avoid interference from nonnumerical alternatives. For that reason we are guilty of the historical bias against number that Capaldi decries.

A notable corrolary to our bias concerns the use of optimal training and testing situations. Perhaps our success in demonstrating transitive inference (Davis, 1992) or ordinal control in rats (Davis & Bradford, 1986), was based on the use of natural test situations involving olfactory cues and simulated burrows. In addition, our procedures in these and other experiments (e.g., Davis, 1984, 1989) involved the use of social reinforcement delivered by an experimenter with whom interaction with the rats had become positive. Similar experimenter-subject rapport may also be fundamental to the success reported by investigators like Boysen and Bernston (1989) and Pepperberg (1987). Each of these experiments appears to maximize the subject's motivation to perform and represents forms of environmental enrichment that, like the previously discussed environmental sterilization of nonnumerical cues, may be necessary for demonstrating numerical competence.

CLEVER HANS AND MISSING CONTROL GROUPS

Having raised the spectre of Clever Hans earlier, it is time to return to that checkered episode that has tempered the way experimental psychologists interact with animal subjects (e.g., Sebeok & Rosenthal, 1981). Rather than bemoaning the tunnel vision of our forefathers, there may be something we can learn from their naivete.

Consider the conclusions we have just reached about how biases can operate on both sides of experimental design: to keep us from appreciating the need for alternative explanations as well as to raise the likelihood of certain alternatives. With that in mind, let us review the original episode of Clever Hans. Why was numerical ability considered such a likely explanation? Why was the possibility of social cuing discounted? What alternative accounts seemed credible at the time?

To begin with, Clever Hans was not the only animal at the turn of the century thought to possess higher mental abilities. As discussed by Rosenthal (1911/1965), Hans's numerical skill must be viewed in a context that includes, to name but a few, Rosa, the mare of Berlin, the Clever dog of Utrecht, the reading pig of London, and Lady, the talking horse of Virginia (who not only read and spelled, but also foretold the future and gave financial advice).

Perhaps the principal difference between Hans and each of these other animals is that Hans's abilities were not displayed by his owner as a source of income. Van Osten, a former mathematics instructor and horse trainer, was a serious and reflective man who was utterly convinced that his animal's abilities were genuine. He allowed an extensive analysis of Hans's performance, which reflected how confident he was that the verdict of inquiry would confirm his belief. Von Osten was described by the eminent psychologist Carl Stumpf (see Rosenthal, 1911/1965) as believing "that the horse is capable of inner speech and thereby enunciates inwardly the number as it proceeds with the tapping" (p. 13).

It is worth reminding ourselves of the kind of problems that Hans routinely solved. Unlike the simpler forms of numerical competence reported in this book, Clever Hans was able to add fractions ("How much is 2/5 and 1/2?") and give his reply (9/10) by separately tapping the numerator and denominator. He was asked the question "What are the factors of 28?" and correctly replied "2, 4, 7, 14, and 28." He was given the number 365,287,419 and asked to insert a decimal after the 8, and report the number in the hundreds place (correct answer, 5). In short, Hans's putative abilities with number eclipse what is reported for the average North American high school graduate barely a century later.

In September 1904, a panel of 13 experts, including Stumpf, as well as a physiologist, a veterinarian, a circus manager, and the director of the Berlin zoo, certified that Clever Hans was demonstrating genuine numerical ability. The panel asserted they could detect no cuing whatsoever, either intentional or unintentional, to account for his performance.

How was this conclusion reached by such worldly and educated observers? That the discovery of a system of social cuing was so long in coming is doubly puzzling because most "clever animals" who performed for their living were known to depend on elaborate forms of cues delivered by their owners or performing partners. Nevertheless, an examination by the "Hans Commission" of 1904 found no such evidence. Stumpf (Rosenthal, 1911/1965) rationalized,

> I did not expect to find the involuntary signals ... in the form of movements. I had in mind rather some form of nasal whisper such as had been invoked by the Danish psychologist A. Lehman in order to explain certain so-called telepathy. I could not believe that a horse could perceive movements which escaped the sharp eyes of the circus manager. . . . One would hardly expect this feat on the part of an animal who was so deficient in keenness of vision. (p. 7)

Stumpf (Rosenthal, 1911/1965) conceded that when the particular cues he anticipated were precluded, he began to consider the possibility that the horse was using number.

> I was ready to change my views with regard to the nature of animal consciousness as soon as a careful examination would show that nothing else could explain the facts, except the assumption of conceptual thinking. I had thought out the process hypothetically; i.e. how one might conceive of the rise of the number concept and arithmetic calculations along the peculiar lines which had been followed in Hans' education. (p. 3)

Stumpf (Rosenthal, 1911/1965) was quick to note that his thinking was in keeping with the *zeitgeist*.

> Zoologists . . . saw in von Osten's results evidence of the essential similarity between human and animal mind, which doctrine has been coming more and more into favor since the time of Darwin. Educators were disposed to be convinced because of the clever systematic method of instruction which had been used and had not, until then, been applied in the education of a horse. (pp. 2–3)

It was not until Stumpf's student, Oskar Pfungst, investigated the situation that the mechanism of Hans's performance was unmasked. Pfungst not only examined the performance of Hans in the horse's normal setting, but took his conclusions back to the laboratory for replication with other horses and human subjects as well. He determined, as is now widely known, that Hans had been cued to stop tapping his foot (the manner in which he "answered" questions) by movements of the questioner's head, eyebrows, or dilation of his nostrils. Pfungst determined that it was not even necessary that such cues be sent by von Osten. Hans was "clever" enough to learn the signaling systems of other interrogators. As long as the questioner knew the correct answer, Hans was also likely to succeed.

Pfungst concluded that the inadvertent delivery of cues resulted from the normal buildup of tension in the questioner as Hans approached the correct answer. Perhaps most telling was the fact that Pfungst, himself, could not stop cuing Hans even after he knew the nature of the signaling system and had attempted to suppress it.

How does the episode of Clever Hans bear on our present attempts to investigate numerical competence in animals? The answer is not a simple admonition against inadvertently cuing our animals. The Clever Hans episode also bears on pronouncements about animal cognition and suggestions about alternative explanations. When we propose a control group or alternative account, we are making a strong statement about the power of an independent variable. The basic test that any control group should pass is whether the control variable would be equally credible as a primary independent variable or explanation. It is constructive to complete the cycle and consider whether the need would exist to control for the original independent variable as an alternative account.

Imagine that von Osten had proclaimed: "I have a horse that is so socially attuned to my presence that the smallest change in my posture or facial muscles, such as twitches in my eyebrow, can serve as signals that cue his behavior. To illustrate this, I will put some relatively complex mathematical questions to the horse. I will ask him, for example, the square root of 16. The horse will "answer" by stomping his foot four times. This will of course be a ruse, since I will have taught him to cease stomping his foot when I show a miniscule facial twitch. Such is the horse's sensitivity to my movements that trained observers, even if they know how "clever animals" are usually trained, may still fail to detect the manner in which he is being cued."

If we were evaluating van Osten's claim today, would someone be likely to exclaim, "Yes, but what if the horse really *knows* the square root of 16?" and demand that we run a control group for numerical competence? Probably not, which may be viewed as evidence for how much more sophisticated we have become about animal cognition. We no longer live in a world where entertainment consists of "reading pigs" or "counting horses." In turn, it no longer seems necessary to control for the possibility that square roots are being computed by a horse. Our search for alternative explanations today might include other forms of external cuing involving visual, auditory, or chemical stimuli, depending on the critic's expertise. Ironically, it is the Clever Hans episode, itself, that has made us more vigilant about scientist-animal interactions in general (e.g., Davis & Balfour, 1992; Sebeok & Rosenthal, 1981).

This exercise underscores several things: First, a control group should offer just as viable an account of the behavior in question as the primary independent variable. Second, our motivation to suggest alternatives is often fueled by conceptual as well as logical dissatisfaction with the original proposal. In making these points, it is possible to draw some instructive parallels between von Osten's counting horse and Capaldi's counting rats.

1. Both Capaldi and von Osten were convinced their animals could count, and despite nearly a century between them, both probably felt that they had an uphill battle in dispelling criticism. It is not clear whose task was more daunting: Van Osten's because of the extent of abilities he wished to claim for Hans, or Capaldi's because the ensuing years had brought a more sophisticated, more entrenched "anti-number" bias.

2. The suggestion of a nonnumerical alternative in both cases met the criterion of viability. In neither case there was what Capaldi and Miller (1988b) termed a "farfetched and fanciful alternative interpretation to counting" (p. 582). In Capaldi's case, as previously noted, the use of "rhythm" had been empirically documented in rats. In von Osten's case, social cuing of trained animals was a well-established phenomenon. Only the fact that cuing could occur both subtly and unintentionally remained to be established.

3. When we follow the exercise of reversing the primary and alternative explanations, as we did in our discussion of the Clever Hans case, there is some difference in the credibility of von Osten's and Capaldi's positions. As noted, most of us today would agree that it is tacitly ludi-

crous to assume that a horse might be computing square roots; thus, control groups would be more reasonably aimed at elucidating the nature of social cuing between human and animal.

In the case of Capaldi's rats, however, the verdict is not so clear. Again following our reversal exercise, if I had used Capaldi and Miller's (1988a) procedure to demonstrate evidence of rhythmic discrimination in rats, would these authors be reasonable in arguing that I run a control group to rule out numerical competence in general or counting in particular? The possibility is by no means as extreme as the Clever Hans example. Yet, is it necessary? Our position has been simply that the ability to count implies a system in which there is considerable sophistication with number (Davis & Memmott, 1982; Davis & Pérusse, 1988). We are not persuaded that rats employ such a system, and we believe the discrimination of regular patterns of event deliveries reflects a *simpler* process. Thus, if rhythm were the primary independent variable, it might be productive to explore its boundaries by controlling for variation in the temporal parameters of delivery, or by varying stimulus modality within and between event sequences to determine whether rhythmic discrimination was amodal. But the need to control for counting or, for that matter, for number, per se, as an alternative account would not be a logical priority.

Finally, there are two aspects of Clever Hans that are often overlooked. First, in our rush to undermine the naivete of numerical claims, we have overlooked the fact that the original demonstration of Clever Hans was quite remarkable in its own right. Mathematics aside, there may not be a clearer demonstration of minimal social cuing between horse and human in the equine literature. Moreover, the Clever Hans episode is often viewed as a dark day in the history of animal experimental psychology. Numerical competence seems to have been most directly affected, although Hans was also reputed to be able to read German and to analyze music. Neither the psychology of reading nor psychomusicology, however, seem to have been damaged by their association with a clever horse. It is time we turned the tables and recognized something positive in the episode. By any reckoning, the work of Pfungst in unravelling the mystery represents a triumph of scientific method over social pressure. Pfungst's work remains credible and rigorous even by contemporary standards (Rosenthal, 1911/1965).

RELATIVE AND ABSOLUTE NUMBER: DICHOTOMY VERSUS CONTINUUM?

An emerging body of literature showing numerical competence in nonhumans suggests that some facility with absolute number is plainly in evidence. Davis and Pérusse (1988) used the term *proto-counting* to describe instances in which investigators have precluded simpler or relative numerical processes, although conclusive evidence for formal enumeration ("true" or "human" counting) is absent.

Subitizing, a much debated perceptual process, remains a potential account for some demonstrations involving the direct apprehension of small absolute quantities. Investigators often take elaborate steps to preclude subitizing on procedural or logical grounds (e.g., Capaldi & Miller, 1988a; Gallistel, 1988). Subitizing is held to be a process akin to pattern matching. When a small number of items are presented in an array, they may be identified in the same way that one recognizes a pattern. In this case, however, the name of the pattern is a number word. To the extent that numbers are small and arrays approximate familiar canonical patterns, recognition time is typically very brief and no formal enumeration is required. In the event that subitizing is the sole method available to nonhumans for processing absolute number, we soon discover it as our tests continue to encroach on the upper limits of the ability (Mandler & Shebo, 1982; von Glasersfeld, 1982).

It is generally held that counting implies ordinality (Gelman & Gallistel, 1978). As Davis and Pérusse (1988) argued, one of the most sensitive ways to demonstrate ordinality is through a procedure called *transitive inference.* Unfortunately, there is scant evidence of logical transivity in nonhuman subjects. Until recently, the only demonstrations were in other primates (see Gillan, 1981; McGonigle & Chalmers, 1977).

Recently, however, we have succeeded in demonstrating transitive inference in rats (Davis, 1992). Our results are based on a procedure in which subjects were trained to construct a series of olfactory cues and correctly infer the "winner" of a novel B versus D comparison from an A B C D E series that had been arbitrarily ordered through a history of reinforced comparisons. These findings represent a major step, not only in demonstrating logical abilities in rats at a level previously associated only with primates, but also in setting the stage for assessing their numerical competence.

The question remains whether relative and absolute numerical skills represent a dichotomy (yielding, perhaps, an "us versus them" in terms of species abilities), or whether there is both a developmental and evolutionary continuum between the two types of abilities. To make the case for a continuum, we must first believe that such a relationship exists in the stimuli themselves. Laurie Hiestand and I are currently investigating this question with rats using a radial maze procedure. We begin with the simple detection of presence versus absence, then require the animal to rank order different amounts of reward on a simple ordinal (i.e., relative) scale (e.g., go to the maze arm containing 12 pellets first, then 6, then 1). Our goal is to test the extent to which additional quantities can be inserted into the series so as to form an interval scale. Presumably, the mechanism for processing an 8 versus 7 choice is different from that involved in a 12 versus 0 or a 12 versus 6 decision (see Thomas, Fowlkes, & Vickery, 1980). Ultimately, differences in how these numerical situations are processed may bear on the continuum between relative and absolute numerical abilities.

Are relative abilities a developmental precursor of absolute numerical skill? Are absolute numerical abilities more readily developed following exposure to relative comparisons within the same range? In short, how does the simplest (and most natural) numerical ability ultimately lead (with help from an experimenter) to those forms of numerical competence displayed by our own species?

CONCLUSIONS

We began by distinguishing between relative and absolute numerosity. According to our position, skill with relative numerosity may be the extent of what is needed by nonhumans in their everyday lives. Absolute numerosity is a human invention, so there is a dearth of related abilities in the animal kingdom. The capacity to formally enumerate or perform operations on numbers is probably confined to humans or, perhaps, the order *primates.*

There is nothing in our thinking or data to rule out the possibility that nonhumans can count. Conversely, there is nothing at present to confirm that they can. If we intend to conclude that rats can count, we need to demonstrate that they possess ordinal knowledge, at least within the range in which enumeration occurs.

Our work on transitive inference has provided yet another piece in the evidential chain required to answer the question of whether rats, or any nonverbal species, can count. We need to remain both vigilant and conservative in confronting the remaining pieces of the puzzle.

ACKNOWLEDGMENTS

Some of the ideas in this chapter were presented in a Symposium on Numerical Competence at the 1988 meeting of the American Psychological Association in Atlanta, GA. The preparation of this chapter was supported in part by a grant to the author from the Natural Sciences and Engineering Research Council of Canada. The author is indebted to Laurie Hiestand, Rachelle Pérusse, and Susan Simmons for their critical comments.

REFERENCES

Boysen, S. T., & Berntson, G. G. (1989). Numerical competence in a chimpanzee *(Pan troglodytes)*. *Journal of Comparative Psychology, 103,* 23–31.
Brackbill, Y., & Fitzgerald, H. E. (1972). Stereotype temporal conditioning in infants. *Psychophysiology, 9,* 569–577.
Capaldi, J. (1988a, August). *Numerical abilities in animals.* Paper presented at American Psychological Association Meeting, Atlanta, GA.
Capaldi, E. J., & Miller, D. J. (1988b). Counting in rats: Its functional significance and the independent cognitive processes that constitute it. *Journal of Experimental Psychology: Animal Behavior Processes, 14,* 3–17.
Capaldi, E. J., & Miller, D. J. (1988c). A different view of numerical processes in animals. *Behavioral and Brain Sciences, 11,* 582–583.
Capaldi, E. J., & Verry, D. R. (1981). Serial anticipation learning in rats: Memory for multiple hedonic events and their order. *Animal Learning & Behavior, 9,* 441–453.
Church, R. M., & Meck, W. H. (1984). The numerical attribute of stimuli. In H. L. Roitbalt, T. G. Bever, & H. S. Terrace (Eds.), *Animal cognition* (pp. 445–464). Hillsdale, NJ: Lawrence Erlbaum Associates.
Davis, H. (1984). Discrimination of the number three by a raccoon *(Procyon lotor)*. *Animal Learning & Behavior, 12,* 409–413.
Davis, H. (1989). Theoretical note on the moral development of rats *(rattus norvegicus)*. *Journal of Comparative Psychology, 103,* 88–90.
Davis, H. (1992). Logical transitivity in animals. In W. K. Honig & G. Fetterman (Eds.), *Cognitive aspects of stimulus control.* Hillsdale, NJ: Lawrence Erlbaum Associates.
Davis, H., & Albert, M. (1986). Numerical discrimination by rats using sequential auditory stimuli. *Animal Learning & Behavior, 14,* 57–59.
Davis, H., & Albert, M. (1987). Failure to transfer or train a numerical discrimination using sequential visual stimuli in rats. *Bulletin of the Psychonomic Society, 25,* 472–474.
Davis, H., Albert, M., & Barron, R. W. (1985). Detection of number or numerousness by human infants. *Science, 228,* 1222.
Davis, H., & Balfour, D. (1992). *The inevitable bond: Examining scientist-animal interactions.* New York: Cambridge University Press.
Davis, H., & Bradford, S. A. (1986). Counting behavior by rats in a simulated natural environment. *Ethology, 73,* 265–280.
Davis, H., & Bradford, S. A. (1987). Simultaneous numerical discrimination by rats. *Bulletin of the Psychonomic Society, 25,* 113–116.
Davis, H., Herrmann, T., & Shattuck, D. (1979). Summation of excitatory and inhibitory control produced by traditional tone-shock contingencies and autocontingencies. *Pavlovian Journal of Biological Science, 14,* 254–262.
Davis, H., & MacFadden, L. (1978). Is autocontingency control established when a traditional contingency is simultaneously available? *Bulletin of the Psychonomic Society, 11,* 387–389.
Davis, H., MacKenzie, K., & Morrison, S. (1989). Numerical discrimination using body and vibrissal touch in the rat. *Journal of Comparative Psychology, 103,* 45–53.
Davis, H., & Memmott, J. (1982). Counting behavior in animals: A critical evaluation. *Psychological Bulletin, 92,* 547–571.
Davis, H., & Memmott, J. (1983). Autocontingencies: Rats count to three to predict safety from shock. *Animal Learning & Behavior, 11,* 95–100.
Davis, H., Memmott, J., & Hurwitz, H. M. B. (1975). Autocontingencies: A model for subtle behavioral control. *Journal of Experimental Psychology: General, 104,* 169–188.
Davis, H., & Pérusse, R. (1988). Numerical competence in animals: Definitional issues, current evidence and a new research agenda. *Behavioral and Brain Sciences, 11,* 561–615.

Davis, H., Shattuck, D., & Wright, J. (1981). Autocontingencies: Factors underlying control of operant baselines by compound tone/shock/shock/no-shock contingencies. *Animal Learning & Behavior, 9,* 322-331.

Gallistel, C. R. (1988). Counting versus subitizing versus the sense of number. *Behavioral and Brain Sciences, 11,* 585-586.

Gelman, R., & Gallistel, C. R. (1978). *The child's understanding of number.* Cambridge, MA: Harvard University Press.

Gillan, D. J. (1981). Reasoning in the chimpanzee: II. Transitive inference. *Journal of Experimental Psychology: Animal Behavior Processes, 7,* 150-164.

Griffin, D. R. (1984). *Animal thinking.* Cambridge, MA: Harvard University Press.

Griffin, D. (1987, June). *How can animal thinking be studied scientifically?* Paper presented at Animal Behavior Society Meeting, Williamstown, MA.

Jaroff, L. (1988, June 13). Fighting against flim flam. *Time Magazine,* pp. 70-72.

Johnson, M. (1988). Out for the count. *Behavioral and Brain Sciences, 11,* 589.

Koehler, O. (1950). The ability of birds to "count." *Bulletin of Animal Behaviour, 9,* 41-45.

Mandler, G., & Shebo, B. J. (1982). Subitizing: An analysis of its component processes. *Journal of Experimental Psychology: General, 11,* 1-22.

McGonigle, B. (1988). Is it the thought that counts? *Behavioral and Brain Sciences, 11,* 593-594.

McGonigle, B. O., & Chalmers, M. (1977). Are monkeys logical? *Nature, 267,* 694-696.

Pepperberg, I. M. (1987). Evidence for conceptual quantitative abilities in the African Grey Parrot: Labelling of cardinal sets. *Ethology, 75,* 37-61.

Pepperberg, I. M. (1988). Studying numerical competence: A trip through linguistic wonderland? *Behavioral and Brain Sciences, 11,* 595-596.

Rosenthal, R. (Ed.). (1965). *Clever Hans.* (Carl L. Rahn, Trans.). New York: Holt, Rinehart & Winston. (Original work published 1911)

Rumbaugh, D. M., Savage-Rumbaugh, E. S., & Hegel, M. T. (1987). Summation in the chimpanzee. *Journal of Experimental Psychology: Animal Behavior Processes, 13,* 107-115.

Rumbaugh, D. M., Savage-Rumbaugh, E. S., & Pate, J. L. (1988). Addendum to summation in the chimpanzee *(Pan troglodytes). Journal of Experimental Psychology: Animal Behavior Processes, 14,* 118-120.

Salman, D. H. (1943). Note on the number conception in animal psychology. *British Journal of Psychology, 33,* 209-219.

Sebeok, T. A., & Rosenthal, R. (Eds.). (1981). *The Clever Hans Phenomenon: Communication with horses, whales, apes, & people.* New York: New York Academy of Sciences.

Thomas, R. K., Fowlkes, D., & Vickery, J. D. (1980). Conceptual numerousness judgments by squirrel monkeys. *American Journal of Psychology, 93,* 247-257.

Von Glasersfeld, E. (1982). Subitizing: The role of figural patterns in the development of numerical concepts. *Archives de Psychologie, 50,* 191-218.

Wesley, F. (1961). The number concept: A phylogenetic review. *Psychological Bulletin, 58,* 420-428.

CHAPTER SIX

Numerical Competence in Animals: A Conservative View

Roger K. Thomas
The University of Georgia

Rosanne B. Lorden
Eastern Kentucky University

About fifty years of work . . . have, in fact, produced a rich crop of several hundred heuristic concepts, but, alas, scarcely a single principle worthy of a place in the list of fundamentals. It is all too clear that the vast majority of the concepts of contemporary psychology, psychiatry, anthropology, sociology, and economics are totally detached from the network of scientific fundamentals. (Bateson, 1972, p. xix)

We assert the conservative views expressed here, because we are concerned that Bateson may have been more right than wrong about psychology and, if so, the reason for the detachment has been the too frequent willingness of psychologists to compromise basic scientific principles. We believe that the integration of the conceptual strengths of cognitive psychology with the methodological strengths of behaviorism is realistic, reasonable, and best for psychological science. It will not help to continue to compromise on behavioral methodology in the interest of preserving "interesting" interpretations of behavior that are not confirmed by the reported observations. In many cases, methodologically superior observations are feasible, and we believe that questionable studies should be redone accordingly. If methodologically appropriate observations are not feasible, then the subject of investigation is not scientific.

Early research on animal cognition was the target of much methodological criticism. Romanes (1883, 1891) was the frequent target of such criticism, and scholars have often cited his works as examples of what not to do when studying the mental abilities of animals. Washburn's

(1926, pp. 4–5) list of objections to the use of anecdotes was directed to Romanes's use of them. Morgan's (1914, p. 53) famous canon against the attribution of higher-order processes to animals when lower-order ones will do is sometimes cited as having been a reaction to Romanes (e.g., Mitchell & Thompson, 1986, p. 6). Such criticism put the study of animal cognition into disrepute and led to its decline as a reputable scientific endeavor. Although Romanes was often the target of methodological criticism, his theoretical views were progressive and we believe he will be increasingly appreciated.

In recent years, there has been a renewal of interest in animal cognition due in part to the rise of "cognitive behaviorism" (Thomas, 1984) or "liberal behaviorism" (Mahoney, 1989) as a reaction against radical behaviorism. Whereas it may have been necessary for such a reaction to occur to save scientific psychology from the stultifying rigidity of radical behaviorism, there are occasional lapses into a liberalization that is too extreme.

For example, Mitchell and Thompson (1986) recently wrote:

> When we use such terms as "think" or "feel" or "want" or "believe" with respect to animals (including humans), we are not necessarily explaining their behavior, but providing names for higher-order descriptive attributes of their behavior. . . . As editors, we admit to being a bit soft on mentalism. (p. xxiv)

In principle, this attitude may be justifiable, but merely to assert it without doing the hard work of establishing the proper use of "think," "feel," and so forth will impede more than help the development of scientific psychology. Some discussion of the difficulties associated with the use of mentalistic terms, especially in the functional sense implied by Mitchell and Thompson, may be found in Thomas and Lorden (1989).

Another example of excessive liberalization is Griffin's (1981) book, *The Question of Animal Awareness,* which Davis (1989) humorously described as "*The Satanic Verses* of animal cognition." Thomas and Lorden (1989) also cited objections to Griffin's book. A third example is Whiten and Byrne's (1988) attempt to justify the use of anecdotal evidence in the study of "deception" by animals. Washburn's (1926) objections to anecdotal evidence are just as applicable today, as Thomas (1988a) and others demonstrated in their commentaries on Whiten and Byrne's article.

THE RETURN OF CLEVER HANS?

Clever Hans was a counting horse whose abilities proved to be fraudulent (see Dewsbury, 1984, p. 189; Rilling, chap. 1, this vol.). Although the fraud may have been inadvertent (as suggested by Pfungst, the psychologist who discovered that Hans was cued by his master), it occurred in an era when reports concerning the mental abilities of animals were already receiving strong scientific opposition. Thus, Clever Hans and animals' use of number became symbolic of the need for rigorous control and theoretical caution before attributing higher-order cognitive processes to animals.

Hence, in important ways, some of the research on animals' use of number, including recent examples, reflects the insidious influence of relaxing methodological standards on the study of animal cognition in general. This chapter addresses some of those methodological compromises and discusses ways more likely to lead to conclusive determinations of numerical competence in animals. Although there are enough methodologically sound studies to conclude that some animals are capable of a conceptual use of number, we show that some recent reviewers have been too uncritical and a more conservative and cautious view of animals' use of number is warranted.

Before proceeding, we should specify that when we use phrases such as "a conceptual use of number," we usually mean the ability to affirm or discriminate the numerousness property of discriminanda on a conceptual basis. The principal exception is counting and, generally, the criteria to demonstrate counting have not been met with animals. What we mean by a "conceptual basis" and the "criteria to demonstrate counting" is discussed later.

REVIEWS OF NUMERICAL COMPETENCE IN ANIMALS

Memories of Clever Hans and early concerns about higher-order processes in animals must have influenced early reviewers of the literature on animals' use of number (e.g., Honigmann, 1942; Salman, 1943; Wesley, 1961) because they were more conservative about what constituted acceptable evidence for an animals' use of number than some recent reviewers. For example, Wesley concluded that only Hicks's (1956) study had been sufficiently free of confounding cues to support the conclusion that an animal had used number (in Hicks's case, the animals were monkeys, *Macaca mulatta,* and the reference number was 3).

Later reviewers, Davis and Memmott (1982), did not cite Wesley (1961) and they failed to explain, for example, why they found acceptable Koehler's research on birds' use of number when Wesley had not. Although Davis and Memmott raised questions concerning the published descriptions of Koehler's work, they also wrote, "The work of Koehler reflects the new rigor of research on higher mental abilities in animals" (p. 551).

Davis and Memmott (1982) also cited favorably and without negative regard an article by Ferster describing chimpanzees *(Pan troglodytes)* use of number (1964; based apparently on research by Ferster & Hammer, 1966). Thomas, Fowlkes, and Vickery (1980) reported that it could not be determined based on the published accounts whether Ferster and Hammer had confounded area with number cues (see related discussion later). Thomas and associates also noted that the number of trials involved (hundreds of thousands!) was consistent with the possibility that the chimpanzees had memorized the specific discriminanda rather than responding to number per se.

More recently, Davis and Pérusse's article (1988a), among its other stated purposes, can be viewed as an updated review that also reexamined much of the older research. They omitted reference to the aforementioned studies by Ferster, cited Wesley (1961) but not in the context of Koehler's work, and reviewed Koehler's work in the same relatively noncritical way as Davis and Memmott (1982).

It will be useful to iterate that Wesley (1961) objected to Koehler's work on the grounds of (a) possible experimenter cues—shades of Clever Hans?—from the manual manipulation of the discriminanda and reinforcers, and (b) the likelihood that odor cues from the food reinforcers guided the subjects' selection of the correct numerical discriminanda.

We do not take the view that Koehler's and Ferster's animals nor the many animals in other studies that could be similarly criticized did not use number cues. According to our view, solutions based on the use of number cues were too often confounded with nonnumerical solutions. The confounded nonnumerical solutions can be eliminated via proper experimental control, so we believe it is in the best long-term interest of the study of animal cognition to disregard confounded studies and replace them with appropriately controlled studies.

Some other possibly confounding cues in animal-number research that have long been recognized but too frequently overlooked or ignored are (a) using items that are uniform in size to construct the numerical discriminanda and, therefore, that confound cumulative area or volume cues with number cues; (b) using uniform backgrounds together with sets of equal area/volume items, permitting differential brightness cues to be used; and (c) using too few patterns of stimuli, especially when the items are figures drawn on cards, permitting the animal to memorize the specific patterns.

There are two recent examples where number and volumetric cues were confounded: Dooley and Gill (1977) used Froot Loops (a proprietary cereal product), which are approximately uniform in size as the numerical items. Rumbaugh, Savage-Rumbaugh, and Hegel (1987) used semisweet chocolate bits of approximately uniform size. We hasten to note that the investigators in both cases acknowledged the confounded cues, and Rumbaugh and colleagues were careful to describe their discriminanda in terms of "quantity" and "amount" rather than number. On the other hand, Rumbaugh and colleagues' stimulus manipulations were described in terms of number, and they used subitizing, a numerical process, to explain their chimpanzees' performances.

In addition to (a) inadvertent experimenter cues, (b) odor cues, (c) area/volume cues, (d) brightness cues, and (e) pattern cues, the possibility of (f) stimulus generalization based on failure to discriminate a new exemplar from a memorized exemplar, and (g) confounding class-concept-use solutions and learning-set-formation solutions faces those who wish to study the use of number concepts by animals. Explications of the latter two problems are summarized later. Before those explications are given and rather than continue citing methodologically inconclusive studies (which considerably outnumber the conclusive ones), it is proposed that any study that claims to have shown the use of number by animals should be examined and found to be free from the aforementioned seven confounded solutions (see Table 6.1).

THE CONCEPTUAL USE OF NUMBER

As noted before, there is no standard definition of concept (Heath, 1967; Kendler & Kendler, 1975; Premack, 1983; Thomas & Crosby, 1977) and, therefore, of using a concept or, in this context, of conceptual behavior.

TABLE 6.1
The Necessary Conditions to Show Animals' Use of Number

I. *Confounds to Be Avoided:*
  1. Inadvertent experimenter cueing.
  2. Odor of reinforcers as discriminative stimulus.
  3. Cumulative area, volume, etc. cues as discriminative stimuli.
  4. Brightness cues based cumulative area etc. cues and uniform background.
  5. Specific pattern memorization.
  6. Stimulus generalization interpretations.
  7. Learning set solutions separable from concept use.

II. *Conditions to Show Conceptual Use of Number:*
  1. Avoid confounds above.
  2. Use trial-unique exemplars.
      OR
  3. Use first-trial data when multitrial problems are used.
      OR
  4. Use text-exemplars without reinforcing responses to them.

Note: See text for explication.

However, it is understood generally that responses to discriminanda based on using a concept must be distinguished from responses based on memorizing specific properties of the discriminanda. Using a concept implies responses *Beyond the Information Given* (so well reflected in the title of a volume of Bruner's works; Anglin, 1973); whereas, responses based on memorizing specific properties implies responses that are limited to the information given.

The conceptual use of information, then, requires that it be free of the possibility of memorizing specific properties. Three ways to preclude responses to exemplars being based on memorizing specific properties of the exemplars are: (a) to present exemplars of the concept one time only; that is, to use trial-unique exemplars; (b) to restrict the critical evidence to the first presentations of exemplars when exemplars are presented more than once, and (c) to present test exemplars without reinforcing the responses to them, so the subject cannot memorize the association between specific exemplars and reinforcement. The latter, an infrequently used procedure, is exemplified in a study by Lombardi, Fachinelli, and Delius (1984), who investigated use of the oddity concept by pigeons. However, their test exemplars were such that a conceptual solution may have been confounded with a stimulus generalization solution based on physical similarities between test and training exemplars (this was suggested by M. R. D'Amato in a personal communication to R. K. Thomas, October 18, 1988).

Conceptualization Versus Stimulus Generalization

Control for stimulus generalization based on physical similarity should be observed in any study of concept use. Although some have equated stimulus generalization with concept formation (e.g., Keller & Schoenfeld, 1950, p. 155; Nevin, 1973, p. 141), we prefer views of stimulus generalization that suggest it is a failure to discriminate (Prokasy & Hall, 1963) or define it in terms of generalization gradients, such as "the degree of generalization will vary inversely with the distance of the stimuli from each other along this dimension. The function expressing this relationship is called the generalization gradient" (Kimble, 1961, p. 484).

Stimulus generalization, as we use the term, implies memorizing a specific reference stimulus or set of reference stimuli followed by responses to new stimuli based on a failure to discriminate between the reference stimulus and the new stimuli. The best way to avoid stimulus generalization in a study that purports to investigate the use of concepts is to preclude the possibility of memorizing a reference stimulus or set of reference stimuli.

6. NUMERICAL COMPETENCE: A CONSERVATIVE VIEW   **133**

Stimulus generalization should be characterized by a curvilinear gradient of affirming responses to exemplars whose physical properties depart systematically from those of some reference stimulus; whereas, concept-use results in a distribution of affirming responses to exemplars of the concept that is essentially rectilinear. The emphasis in both cases is on affirmation of exemplars as opposed to distributions of response times, and so forth. Obviously, the difference between stimulus generalization and conceptualization is not as clear as implied here and more could be said about them. However, we must move on.

Using Class Concepts Versus Learning Set Solutions

Many reports of animals' use of concepts have been based on transfer of training without using trial-unique exemplars, first-trial evidence, or nonreinforced test exemplars. Perhaps, the ideal example of transfer of training based on memorizing is represented in object quality learning set formation where better-than-chance performances on the second trial of new problem presentations constitute the best evidence of learning set formation (e.g., Harlow, 1949, 1959; Hodos, 1970). We do not suggest that learning set per se is not conceptual (see Thomas, 1989) but that responses based on using class concepts and responses based on object quality learning set formation are empirically and logically distinguishable.

Thomas and Noble (1988) showed that using the oddity concept (choosing the discriminandum that had the odd odor among three discriminanda, two of which had the same odor) was distinguishable from learning set formation. Following pretraining that was closely related to the oddity learning set training, their rats performed much better than chance on Trial 2 early within a series of 300, five-trial problems, but the rats never performed better than chance on Trial 1. The odd odor is evident on Trial 1, thus it was possible to use the oddity concept and respond better than chance on Trial 1. That they did not respond better than chance on Trial 1 but did on Trial 2 indicated they learned to identify the correct stimulus on Trial 1 (presumably via the well-known "win-stay/lose-shift" strategy; e.g., Levine, 1965) and respond to it based on rote memory on Trials 2–5.

Aversion to new stimuli likely did not account for the rats' poor performances on Trial 1, because (a) they responded unhesitatingly on Trial 1 when both the odd and nonodd stimuli were new, (b) only 16 odors were used and two were used per problem; thus, early in training the uniqueness of the problems was based on the combinations of odors rather than new odors, and (c) the odors were food flavorings to which one might expect rats to be attracted. Thomas and Noble's study shows that any report of an animal's use of a concept that fails to distinguish between Trial 1 and Trial 2 performances has the potential of confounding responses based on use of the concept with responses based on learning-set-formation in conjunction with rote-memory solutions.

THE PROCESSES FOR NUMERICAL DETERMINATIONS

In the present context, perhaps the most important paper in terms of describing processes of visual number judgments was that of Kaufman, Lord, Reese, and Volkmann (1949). They discussed three processes: (a) *subitizing,* a term they coined, (b) *estimating,* and (c) *counting.* Subitizing and estimating were distinguished by four findings that emerged from their study of humans' abilities to determine the number of dots in random arrays presented for 200 ms. They found that arrays containing from 1–6 dots were determined more (a) accurately, (b) confidently, and (c) rapidly than arrays of more than 6 dots. (d) A "breakpoint" in the data that occurred between 6 and 7 dots was also used to distinguish between subitizing and estimation. Counting, the third process acknowledged by Kaufman and associates, was presumed to have been precluded by the 200 ms presentation times.

Animal studies to date have not addressed confidence nor rapidity of response, and accuracy and the "breakpoint" between 6 and 7 have not been addressed in the sense that Kaufman and associates (1949) did. For that matter, accuracy has been assessed only in the sense of animals being able to discriminate between successive arrays of discriminanda (e.g., 3 vs. 4, 7 vs. 8, etc.; hereafter, such discriminations are abbreviated as 3:4, 7:8, etc.) and the "breakpoint" between 6 and 7 has not been addressed systematically in any way.

In fact, in the best controlled animal studies to date that examined successive arrays with 6 or more items, the "breakpoint" appears to occur between 8 and 9 or between 9 and 10. For example, both of the squirrel monkeys used by Thomas and colleagues (1980; *Saimiri sciureus sciureus*) met a stringent criterion (45 correct in a 50-trial session) discriminating between two simultaneously present arrays of 7:8 "dots" (using trial-unique problems controlled for area, brightness, and specific pattern cues). One of the two monkeys met criterion on 8:9 dots in the 500 trials allowed but failed to reach criterion on the 9:10 problem.

Terrell and Thomas (1990) found similar limits using the *number of sides* (or angles) of quasi-randomly constructed polygons as discriminanda; they also used trial-unique problems and controlled for area, brightness, and specific pattern cues. Two of four monkeys (one was *Saimiri scuireus sciureus;* the other, *S. boliviensus boliviensus*) met rigorous criteria (36 correct in a 40-trial session) distinguishing between heptagons and octagons but failed to reach criterion on the octagon versus nonagon problem in the 1,000 trials allowed.

Subsequent to Kaufman and associates (1949), investigations of "subitizing" in humans have raised questions about its definition and distinguishing criteria, including whether counting had been precluded by the 200 millisecond presentation times (e.g., Folk, Egeth, & Kwak, 1988; Mandler & Shebo, 1982). For reasons such as these, Terrell and Thomas (1990) suggested that subitizing as a term and hypothesized process had outlived its usefulness. It should be noted also that subitizing has never been an acceptable explanatory term, rather it is a descriptive term that was meant to reflect the direct apprehension of number without explaining how that might have occurred.

Davis and Pérusse (1988a) offered the most recent account of possible processes of numerical competence in animals. Table 6.2 includes a summary of their list of processes of numerical competence. Table 6.2 also includes our shorter but equally comprehensive list of processes (in terms of what animals are likely to use). Justifications for our disagreement with Davis and Pérusse's list follow.

Davis and Pérusse's list retained Kaufman and associates' (1949) subitizing, estimating, and counting and added "relative numerousness judgments" as the major processes. Subsumed under counting, Davis and Pérusse added "protocounting," which was defined as:

> Instances in which counting has been identified as the most likely numerical process (e.g., in situations where relative numerousness judgments and subitizing have been precluded), although control tests (e.g., for transfer) have not revealed evidence of true counting. (p. 562)

Davis and Pérusse also subsumed "Concept of Number" under counting, described it as "an attribute of true counting," and defined it "in terms of abstract or modality-free numerical ability and revealed in the capacity to transfer numerical discriminations across sense modalities or procedures" (p. 562).

In his commentary on Davis and Pérusse (1988a), Thomas (1988b, p. 600) argued that protocounting was "unjustified and unjustifiable" by (a) asking how one could ever know that other processes had been precluded and (b) noting that the last clause in the previous quotation, which defined protocounting, made no sense methodologically (one is required to prove the absence of "true counting," which is tantamount to proving the null hypothesis).

We also disagree with the necessity to use cross-modal or cross-procedural transfer to demonstrate "Concept of Number," as Davis and Pérusse (1988a) suggested. If an animal can demonstrate use of number

TABLE 6.2
A New Glossary of Processes of Numerical Competence:
Supersedes Davis and Pérusse's (D&P) Glossary (1988a).

The New Glossary:
1. *Prototype matching (PM)*.

	Analogous to Rosch's (1975) use of prototype, and applied to numerousness discriminanda.
A. *Absolute PM*	May be precise, e.g., "sevenness" or imprecise, e.g., "manyness."
B. *Relative PM*	May be precise, e.g., "fewer" when applied to successive sets of entities (e.g., 7 vs. 8) or imprecise, e.g., "fewer" when applied to nonsuccessive sets (e.g., 25 vs. 50).
2. *Counting*	We agree with D&P on using at least the first three principles of counting presented by Gelman and Gallistel (1978; G&G) as the defining, operational criteria for counting. However, unlike D&P and G&G, we deem it to be necessary to show evidence for acquisition of a symbol system and the use of "tagging."

Davis and Pérusse's Glossary:
1. *Relative numerousness judgments*

	Subsumed by relative prototype matching.
2. *Subitizing*	Has outlived usefulness due to (a) loss of original meaning, (b) being descriptive but not explanatory, and (c) being replaceable by descriptive and explanatory process, prototype matching.
3. *Estimation*	Retains limited use but not in sense used by D&P. Their use is replaceable with prototype matching as previously indicated.
4. *Counting*	See *Counting* in *The New Glossary* in this Table.
A. *Protocounting*	Unnecessary, unjustified, and as D&P defined it, unjustifiable.
B. *Concept of number*	D&P's use is both too restrictive—need for cross-modal and cross-procedural transfer tests—and too unrestrictive—conditions set forth in Table 6.1 here.

Note: The new glossary is believed to be exhaustive in terms of the basic processes of numerical competence that animals are likely to use.

when the seven confounding conditions described earlier have been excluded and, especially, when trial-unique discriminanda are used, we believe it is appropriate to say that the animal is using number conceptually. However, we acknowledge this may enter the realm of discussion on the question, "What is the difference between a concept and a percept?", which is best left for another time. Intramodal and intraprocedural demonstrations of numerical determinations with trial-unique discriminanda are consistent with the larger body of literature on human and nonhuman animals' use of class concepts, whether they are based on color, shape, size, number, or combinations thereof. We do agree that a stronger case is made for more abstract conceptual numerical determinations when transfer is demonstrated with cross-modal or cross-procedural designs.

6. NUMERICAL COMPETENCE: A CONSERVATIVE VIEW   137

To summarize, Davis and Pérusse (1988a) identified four major processes of numerical competence: relative numerousness judgments, subitizing, estimation, and counting; protocounting, a process subsumed under counting was a fifth process. Our position, as discussed earlier, mandates that *subitizing* as a term and process should be abandoned and *protocounting* is unjustified and, as they defined it, unjustifiable. We agree with Davis and Pérusse that the first three of Gelman and Gallistel's (1978) five principles of counting may provide the best definition of and criteria for counting. Remaining to be considered is the usefulness of the processes implied by relative numerousness judgments and estimation.

Before we can give proper consideration to relative numerousness judgments and estimation, we need to say a bit more about counting and to introduce in the context of number judgments the process of prototype matching. We then suggest that prototype matching is both the underlying process for estimation and a prerequisite for relative numerousness judgments.

Counting

As noted, Gelman and Gallistel's (1978, pp. 77–82) principles of counting, especially the first three, provide the best definition and criteria for counting. Briefly, these are: (a) the one-to-one principle, refers to "tagging" each item uniquely; (b) the stable-order principle, which means the tags must correspond to the items in a "stable, repeatable order"; note that this refers to the tags and not to the items, for example, you may count the stars in the "Big Dipper" from any starting and ending position but you must apply the tags, one per star, in the same order; (c) the cardinal principle, which means the last tag applied to the last item describes the number of items in the set; (d) the abstraction principle, which means that if one can count, one can count any set of items; and (e) the order-irrelevance principle as applied to the items to be counted; see note in (b). That the fourth and fifth principles are not essential to the evidence for counting has been discussed by Gallistel (1990, pp. 339–340) and Thomas (in press), and Davis and Pérusse (1988a) also questioned their necessity.

We note, however, as did Davis and Pérusse (1988a, p. 562) that other definitions of counting have been more lenient in terms of the evidence implied in Gelman and Gallistel's principles. We disagree with these more lenient definitions but do not take time to argue the case here other than to agree with Davis and Pérusse that data obtained in accordance with such definitions might be explained by noncounting processes.

We disagree strongly with Davis and Pérusse (1988a) and with Gelman and Gallistel (1978) that one need not show observable evidence of tagging. For example, Gelman and Gallistel wrote: "We hold open the possibility that animals . . . may tick off items in an array, one by one, with distinct mental tags employed in a fixed order, and use the final mental tag as a representation of numerosity" (p. 77). Davis and Pérusse wrote, "verbal number tags need not be replaced by overt physical markers" (p. 566).

Our disagreement with Gelman and Gallistel (1978) and Davis and Pérusse (1988a) concerning the need for observable evidence of tagging and, more importantly, evidence of its underlying symbol system is based on the following three points:

1. Many instances of alleged counting by animals might be explained by noncounting processes; see Davis and Pérusse (1988a), who cited several examples with which we agree. Thus, in the absence of evidence for tagging and its underlying symbol system, counting may be mistakenly inferred when a noncounting process was used.

2. The use of tags implies the use of an underlying symbol system and the acceptance of "mental tags" or the absence of "overt physical markers" omits the need to show the evidence that the animal had an opportunity or ability to acquire the symbol system. We believe that after experience, the symbols might become internalized and used in the absence of overt physical markers, but one should be required at some point to show evidence for the acquisition of the symbol system. We believe that most claims of animal counting are questionable, because there was no opportunity in the animal's reported experiences to have acquired the necessary symbol system.

3. Ifrah (1985) reported that some numerically underdeveloped human cultures that apparently lacked the ability to count, nevertheless had precise number usage up to 4. Beyond 4, number usage in such cultures was reflected in terms analogous to "many" or "countless." Generally, number use greater than 4 involved a need for explicit, physical substitution (symbolic) of tags (e.g., body parts, notched sticks, pebbles or beads, etc.). It is our view that if humans who lack a linguistic system for counting must use physical tags in order to count numbers greater than 4, it is likely that nonhuman animals must also.

In this regard, we note that in the best study of animal counting to date, Boysen and Berntson's (1989) chimpanzee *(Pan troglodytes)*, Sheba, was trained at the time of their report to count up to 4. There was strong evidence to show that Sheba had acquired a symbol system, she had a knowledge of ordinality, and she gave some signs of partitioning and tagging. However, in view of evidence that noncounting humans can use numbers precisely up to 4 without counting, it will be important to show that animals such as Sheba can count numbers greater than 4.

Prototype Matching

For the remainder of this chapter, it is assumed that the processes discussed apply to the discrimination of number when counting has not been or cannot be demonstrated.

Rumbaugh and associates (1987) seemed to express a perceptual view of certain number discriminations when, in reference to Hicks's study (1956) showing rhesus monkeys use of "threeness," they wrote:

> We suggest that a perceptual norm, rather than a number concept, has been acquired and that such a norm is based on the limited perceptual configurations inherent in the simultaneous presentation of three items or objects. (p. 108)

It is clear from the text following the aforementioned quotation that "perceptual norm" applied to the relatively few patterns that a "small number of things" affords. However, the chapter by Rumbaugh in this volume suggests a broader implication comparable to that of prototype matching as used in this chapter.

Davis and Pérusse (1988b) also provided a strong hint of a perceptual process (without naming it, although they spoke in terms of "identify[ing] against a template," p. 604) in their response to Gallistel's (1988) commentary on Davis and Pérusse (1988a). Gallistel cited reaction time data from Mandler and Shebo's (1982) study of "subitizing" in humans to make the following argument:

> Thus, when dot arrays are presented in fixed *Gestalten* (like those on the faces of a die), then the reaction time for judging their numerosity is indeed what one would expect if twoness and threeness were directly perceptible attributes like cowness and treeness, but when the arrays do not have a fixed pattern, then the reaction time function is what you would expect from a counting process. (p. 586)

Davis and Pérusse (1988b) opposed Gallistel's argument as one for counting with the following:

> We also believe that different representations of cows might well occasion a continuum of reaction times to correctly identify them as "cows." A conventional (canonical) photograph of a Holstein cow might yield a reaction time that is considerably shorter than, perhaps, a pen-and-ink caricature or, more tellingly, a cubist rendering. . . . Does it suggest that the receiver had to resort to enumeration to label the more extreme cows correctly? We believe not. (p. 604)

We agree with their opposition to Gallistel's argument in support of a counting interpretation. However, Davis and Pérusse used the previous quotation to argue for a subitizing interpretation, and our objections to subitizing have been noted.

Von Glasersfeld (1982) viewed subitizing as a perceptual process involving empirical abstraction to attend

> not to the specific content of experience, but to the operations that combine perceptual and proprioceptive elements into more or less stable patterns. These patterns are constituted by motion, either physical or attentional, forming "scan paths" that link particles of sensory experience. To be actualized in perception or representation, the patterns need sensory material of some kind, but it is the motion, not the specific sensory material used, that determines the pattern's character. Because of the dependence on some (unspecified) sensory material and motion, they are called figural patterns. (p. 196)

The formation of such figural patterns may involve spatial or temporal configurations of perceptual items. In young children, these figural patterns (e.g., a specific pattern of dots on a die) are associated with number words by a semantic connection and not because of the number of perceptual units of which they are composed. In animals, such figural patterns may be associated with reinforcement. The ability to form an association between a figural pattern on a die and a number word does not require a concept of number, nor does the ability to associate certain figural patterns with reinforcement mean an animal is counting. Such figural patterns have numerosity, but a subject's discriminative response to them may or may not reflect a concept of number depending on whether the conditions in Table 6.1 have been met.

What is the nature of these figural patterns and how may they be used in number discrimination tasks? Davis and Pérusse (1988b) suggested the possibility of comparing to a template. Presumably, with such a model subjects would form a template for each figural pattern involving a different number of perceptual elements. A stimulus, then, would be compared against a set of templates that have been stored in memory. The stimulus is compared to these templates in memory until a match or near-match is found. Such a hypothesized process has been found to be inadequate as a basis for pattern recognition because of its inflexibility, the infinity of templates required to recognize all possible figural patterns, and the temporal inefficiency of such a process (Matlin, 1989).

Terrell and Thomas (1990) noted the relationship between Gallistel's (1988) "cowness" versus numerousness example, Davis and Pérusse's (1988b) counterargument quoted earlier, and prototype matching interpretations of using concepts as exemplified in the work of Rosch and her colleagues. Prototypes are abstract, idealized patterns that are stored in memory. When we see a new stimulus, we compare it to a stored prototype. The match does not have to be exact. In this sense, a prototype is a construction stored in memory of an average or representative pattern. The construction of a prototype is based on experience with many exemplars of a concept, and the prototype becomes a representation of the concept.

Rosch (1975) scaled a number of conceptual categories, including "bird" (but not "cow"), in terms of the closeness of exemplars to a prototype, and reaction time measures were included among others to scale the exemplars. The American robin was deemed to be an ideal exemplar of bird and was given a scale index of 1; other exemplars among the 54 birds listed and their scaled values were raven (2.01), pelican (2.98), and chicken (4.02). A stimulus that is similar to its prototype will usually require less time to match and result in a faster response.

Prototype matching is a well-established process to explain the acquisition and use of class concepts in general, and we suggest that numerousness concepts are not an exception. Although our emphasis is on simultaneously present sets of items, within limits, the prototype matching process should be applicable to temporally spaced sets of entities or events as well.

Terrell and Thomas (1990) suggested that for simultaneously present arrays of items, animals may acquire and use prototypes such as "twoness," "sevenness," and so forth. They proposed that prototype matching could account for data previously subsumed under subitizing because: (a) Prototype matching, unlike subitizing, is not defined in terms of measures such as accuracy, rapidity, and confidence, which are usually not assessed; and (b) prototype matching is both descriptive and explanatory whereas subitizing is only descriptive, so it was suggested that prototype matching replace subitizing.

Prototype Matching and Estimation

As noted earlier, we suggest that estimation is based on prototype matching. *Estimation* is a general term usually associated with approximate determinations of quantitative values, such as: How many? How heavy? How tall? How much time?. Accurate estimates in categories such as these rely on prior experience estimating number, weight, length, and time, and such experience is the basis for the formation of prototypes. The amount and kind of meaningful experience determines the strength of a prototype and, therefore, the reliability and validity of an estimate.

Estimates can in some instances be reliable and precise and in other instances reliable and imprecise. Precision is reflected in animals' abilities to discriminate between arrays constructed of successive numbers of items. The highly experienced monkeys in the Thomas and associates (1980) and Thomas and Terrell (1990) studies distinguished between sevenness and eightness reliably and precisely. However, the general failure of the monkeys to distinguish eightness from nineness suggests that there is a limit to precise numerousness estimates and it may be impossible to form precise prototypes that enable one to discriminate between successive random arrays of items numbering eight or more. If the latter proves to be the case, it may be one more instance of the limits of information capacity as discussed in Miller's (1956) well-known article, "The Magical Number Seven, Plus or Minus Two: Some Limits on Our Capacity to Process Information."

On the other hand, it is likely that monkeys could form discriminable but imprecise prototypes for nonsuccessive arrays involving more than eight items. For example, one can readily imagine that monkeys could learn, for example, to discriminate between random arrays of 25 and 50 dots. One can also imagine that monkeys could learn to recognize "twenty-fiveness" reliably using a discrimination procedure as long as the contrasting arrays differed sufficiently from 25. No doubt, there are systematic psychophysical difference thresholds associated with the discrimination of numerical arrays, and it will be of interest to investigate such in animals.

Prototype Matching and Absolute and Relative Numerousness Judgments

It is first useful to distinguish between absolute and relative numerousness judgments. This distinction is related directly to the distinction between absolute and relative class concepts proposed by Thomas and Crosby (1977) and discussed in the context of number by Thomas and associates (1980). In fact, numerousness concepts are examples of class concepts. Exemplars associated with absolute numerousness concepts, like exemplars of all absolute class concepts, have their concept-affirming properties among the inherent properties of each exemplar. For example, threeness is an inherent property of three simultaneously present items. Exemplars of relative numerousness concepts, such as "more" and "fewer" are not inherent in the exemplar; for example, 7 items are "more" when contrasted with 6 items but 7 items are "fewer" when contrasted with 8 items. A simple operational criterion distinguishes between absolute and relative numerousness (or other class) concepts; namely, whether it is necessary to compare exemplars in order to affirm the one that manifests the concept. It is necessary to compare exemplars of relative numerousness concepts but not of absolute numerousness concepts.

6. NUMERICAL COMPETENCE: A CONSERVATIVE VIEW   143

In the absence of counting, both absolute and relative numerousness judgments likely depend on prototype matching. A strong hint of this likelihood was seen in the Thomas and associates (1980) study. One of the two monkeys showed immediate transfer (responding correctly on 46 of 50 trials in the first session) to a 6:7 discrimination in the session following his having met criterion on a 5:6 discrimination. Responses to "fewer" had been reinforced throughout, which meant that the reinforcement contingency associated with sixness was reversed between the 5:6 and 6:7 problems. Prior to the 6:7 problem, "sixness" and "sevenness" had each been associated with nonreinforcement on a total of four previous problems (viz., 2:6, 3:6, 4:6, 5:6, 2:7, 3:7, 4:7, and 5:7), but responses to sixness or sevenness per se had never been reinforced prior to the 6:7 problem. Presumably the extensive experience with sixness and sevenness in problems prior to the 6:7 problem had resulted in the monkeys' acquiring discriminable prototypes for sixness and sevenness, which enabled the immediate discrimination of the 6:7 dot arrays. Consistent reinforcement for choosing "fewer" in several prior problems enabled the immediate selection of sixness on the 6:7 problem.

It may be noted that Thomas and associates (1980) deliberately confounded the absolute and relative class concept solutions in order to maximize the possibility of determining the monkeys best discriminative performances. Examples of studies that controlled against an absolute class concept solution and, therefore, showed a relative class concept solution are those of Dooley and Gill (1977) and Thomas and Chase (1980). Dooley and Gill demonstrated a chimpanzee's *(Pan troglodytes)* use of "more," and "less," and Thomas and Chase investigated squirrel monkeys' *(Saimiri sciureus sciureus)* use of "more," "less," and "intermediate" numerousness. The statistically reliable responses of the monkeys to "intermediate" also shows that they were capable of *ordinal judgments* of numerousness.

CONCLUSIONS

In conclusion, if our arguments are accepted (a) that subitizing and protocounting as number judgment processes should be abandoned and (b) that prototype matching is the basis for absolute and relative numerousness judgments, then prototype matching and counting are the only processes needed to account for numerical competence by animals.

A demonstration of the use of number necessarily implies a conceptual use of number. Otherwise, the results can be explained by the use of nonconceptual cues or nonnumerical solutions, especially those based on memorizing specific properties or patterns associated with numerousness discriminanda. In order to eliminate interpretations based on non-conceptual cues and solutions, it is necessary to control against the use of (a) inadvertent experimenter cues, (b) odor from the reinforcers, (c) confounded area or volume, (d) differential brightness cues, and (e) specific pattern cues, (f) stimulus generalization as opposed to conceptualization, and (g) rote-memory/learning-set-based solutions. In order to demonstrate a conceptual use of number, it is necessary to use (a) trial-unique exemplars, (b) only first-trial data when multitrial problems are used, or (c) data obtained from responses to test-exemplars that have not been associated with reinforcement; in all three cases, a stimulus generalization interpretation as summarized here must be excluded.

Regarding the processes by which number is used, counting is of the highest order, and we believe that Gelman and Gallistel's (1978) first three principles of counting provide the best definition and criteria of the evidence necessary to show counting. However, we disagree strongly with their assertion that it is not necessary to demonstrate a subject's use of tags. It is necessary at some point to demonstrate the use of tagging, because it must be shown that the subject has acquired the prerequisite symbol system that is the basis for tagging and counting.

Regarding noncounting processes, the most important numerical process is prototype matching. Prototype matching is a general process that describes and explains how humans and other animals use class concepts in general and, for present purposes, numerousness concepts in particular. Prototype matching is the basis for absolute numerousness judgments (e.g., affirming the "sevenness" of a set of seven items) and is a prerequisite for relative numerousness judgment (e.g., "more" and "fewer"). Prototype matching can be precise, such as, discriminating between arrays defined by successive numbers or imprecise, such as, discriminating between nonsuccessive arrays. The term *estimation* remains useful, because it is a general term related to quantitative judgments including but not limited to numerousness. Estimation is a descriptive, not an explanatory, term, and the basis for estimation is prototype matching, whether the estimated quantity involves length, area, volume, mass, time, number, and so forth.

REFERENCES

Anglin, J. M. (Ed.). (1973). *Beyond the information given.* New York: W. W. Norton.
Bateson, G. (1972). *Steps to an ecology of mind.* New York: Ballantine.
Boysen, S. T., & Berntson, G. G. (1989). Numerical competence in a chimpanzee *(Pan troglodytes). Journal of Comparative Psychology, 103,* 23–31.
Davis, H. (1989, June). *Anthropomorphism reconsidered: Animal cognition vs. animal thinking.* Paper presented at the meeting of the Animal Behavior Society, Highland Heights, KY.

Davis, H., & Memmott, J. (1982). Counting behavior in animals: A critical review. *Psychological Bulletin, 92,* 547–571.
Davis, H., & Pérusse, R. (1988a). Numerical competence in animals: Definitional issues, current evidence, and a new research agenda. *Behavioral and Brain Sciences, 11,* 561–579.
Davis, H., & Pérusse, R. (1988b). Numerical competence: From backwater to mainstream of comparative psychology. *Behavioral and Brain Sciences, 11,* 602–615.
Dewsbury, D. A. (1984). *Comparative psychology in the twentieth century.* Stroudsburg, PA: Hutchinson Ross.
Dooley, G. B., & Gill, T. V. (1977). Acquisition and use of mathematical skills by a linguistic chimpanzee. In D. M. Rumbaugh (Ed.), *Language learning by a chimpanzee: The Lana project* (pp. 247–260). New York: Academic Press.
Ferster, C. B. (1964). Arithmetic behavior in chimpanzees. *Scientific American, 210,* 98–106.
Ferster, C. B., & Hammer, Jr., C. E. (1966). Synthesizing the components of arithmetic behavior. In W. K. Honig (Ed.), *Operant behavior: Areas of research and application* (pp. 634–676). New York: Appleton-Century-Crofts.
Folk, C. L., Egeth, H., & Kwak, H-W. (1988). Subitizing: Direct apprehension or serial processing? *Perception & Psychophysics, 44,* 313–320.
Gallistel, C. R. (1988). Counting versus subitizing versus the sense of number. *Behavioral and Brain Sciences, 11,* 585–586.
Gallistel, C. R. (1990). *The organization of learning.* Cambridge, MA: MIT Press.
Gelman, R., & Gallistel, C. R. (1978). *The child's understanding of number.* Cambridge, MA: Harvard University Press.
Griffin, D. R. (1981). *The question of animal awareness* (2nd ed.). New York: Rockefeller University Press.
Harlow, H. F. (1949). The formation of learning sets. *Psychological Review, 56,* 51–65.
Harlow, H. F. (1959). Learning set and error factor theory. In S. Koch (Ed.), *Psychology: A study of a science: Vol. 2. General systematic formulations, learning and special processes* (pp. 492–537). New York: McGraw-Hill.
Heath, P. L. (1967). Concept. *The encyclopedia of philosophy* (Vol. 2). New York: Free Press.
Hicks, L. H. (1956). An analysis of number-concept formation in the rhesus monkey. *Journal of Comparative and Physiological Psychology, 49,* 212–218.
Hodos, W. (1970). Evolutionary interpretation of neural and behavioral studies of living vertebrates. In F. O. Schmitt (Ed.), *The neurosciences: Second study program* (pp. 26–39). New York: Rockefeller University Press.
Honigmann, H. (1942). The number conception in animal psychology. *Biological Review, 17,* 315–337.
Ifrah, G. (1985). *From one to zero: A universal history of numbers* (L. Bair, Trans.). New York: Viking Penguin. (Original work published 1981)
Kaufman, E. L., Lord, M. W., Reese, T. W., & Volkmann, J. (1949). The discrimination of visual number. *American Journal of Psychology, 62,* 498–525.
Keller, F. S., & Schoenfeld, W. N. (1950). *Principles of psychology.* New York: Appleton-Century-Crofts.
Kendler, H. H., & Kendler, T. S. (1975). From discrimination learning to cognitive development: A neobehavioristic odyssey. In W. K. Estes (Ed.), *Handbook of learning and cognitive processes: Vol. 1. Introduction to concepts and issues* (pp. 191–247). New York: Wiley.
Kimble, G. A. (1961). *Hilgard and Marquis' conditioning and learning.* New York: Appleton-Century-Crofts.
Levine, M. (1965). Hypothesis behavior. In A. M. Schrier, H. F. Harlow, & F. Stollnitz (Eds.), *Behavior of nonhuman primates* (Vol. 1, pp. 97–127). New York: Academic Press.

Lombardi, C. M., Fachinelli, C. C., & Delius, J. D. (1984). Oddity of visual patterns conceptualized by pigeons. *Animal Learning & Behavior, 12,* 2–6.

Mahoney, M. J. (1989). Scientific psychology and radical behaviorism: Important distinctions based in scientism and objectivism. *American Psychologist, 44,* 1372–1377.

Mandler, G., & Shebo, B. J. (1982). Subitizing: An analysis of its component processes. *Journal of Experimental Psychology: General, 111,* 1–22.

Matlin, M. (1989). *Cognition* (3rd ed.). New York: Holt, Rinehart & Winston.

Miller, G. A. (1956). The magical number seven, plus or minus two: Some limits on our capacity to process information. *Psychological Review, 63,* 81–97.

Mitchell, R. W., & Thompson, N. S. (1986). *Deception: Perspectives on human and nonhuman deceit.* Albany, NY: State University of New York Press.

Morgan, C. L. (1914). *Introduction to comparative psychology* (rev. new ed.). New York: Scribner's.

Nevin, J. A. (1973). Stimulus control. In J. A. Nevin & G. S. Reynolds (Eds.), *The study of behavior* (pp. 114–152). Glenview, IL: Scott, Foresman.

Premack, D. (1983). Animal cognition. *Annual Review of Psychology, 34,* 351–362.

Prokasy, W. F., & Hall, J. F. (1963). Primary stimulus generalization. *Psychological Review, 70,* 310–322.

Romanes, G. J. (1883). *Animal intelligence.* New York: D. Appleton.

Romanes, G. J. (1891). *Mental evolution in animals.* New York: D. Appleton.

Rosch, E. (1975). Cognitive representations semantic categories. *Journal of Experimental Psychology: General, 104,* 192–233.

Rumbaugh, D. M., Savage-Rumbaugh, S., & Hegel, M. T. (1987). Summation in the chimpanzee *(Pan trogolodytes). Journal of Experimental Psychology: Animal Behavior Processes, 13,* 107–115.

Salman, D. H. (1943). Note on the number conception in animal psychology. *British Journal of Psychology, 33,* 209–219.

Terrell, D. F., & Thomas, R. K. (1990). Number-related discrimination and summation by squirrel monkeys (*Saimiri sciureus sciureus* and *Saimiri boliviensus boliviensus*) on the basis of the number of sides of polygons. *Journal of Comparative Psychology, 104,* 238–247.

Thomas, R. K. (1984). Are radical and cognitive behaviorism incompatible? *The Behavioral and Brain Sciences, 7,* 650–651.

Thomas, R. K. (1988a). Misdescription and misuse of anecdotes and mental state concepts. *The Behavioral and Brain Sciences, 11,* 265–266.

Thomas, R. K. (1988b). To honor Davis and Pérusse and repeal their glossary of numerical competence. *Behavioral and Brain Sciences, 11,* 600.

Thomas, R. K. (1989, March). *Conceptual behavior and learning set formation.* Paper presented at the meeting of the Southern Society for Philosophy and Psychology, New Orleans, LA.

Thomas, R. K. (in press). Primates' conceptual use of number: Ecological perspectives and psychological processes. In T. Nishida, W. C. McGrew, P. Marler, M. Pickford, & F. B. M. de Waal (Eds.), *Proceedings of the XIIIth Congress of the International Primatological Society, Vol. 1, Human Origins.* Tokyo: Tokyo University Press.

Thomas, R. K., & Chase, L. (1980). Relative numerousness judgments by squirrel monkeys. *Bulletin of the Psychonomic Society, 16,* 79–82.

Thomas, R. K., & Crosby, T. N. (1977). Absolute and relative class conceptual behavior in squirrel monkeys *(Saimiri sciureus). Animal Learning & Behavior, 5,* 265–271.

Thomas, R. K., Fowlkes, D., & Vickery, J. D. (1980). Conceptual numerousness judgments by squirrel monkeys. *American Journal of Psychology, 93,* 247–257.

Thomas, R. K., & Lorden, R. B. (1989). What is psychological well being? Can we know if primates have it? In E. F. Segal (Ed.), *Housing, care and psychological wellbeing of captive and laboratory primates* (pp. 12–26). Park Ridge, NJ: Noyes Publications.

Thomas, R. K., & Noble, L. M. (1988). Visual and olfactory learning in rats: What evidence is necessary to show conceptual behavior? *Animal Learning & Behavior, 16,* 157–163.
von Glasersfeld, E. (1982). Subitizing: The role of figural patterns in the development of numerical concepts. *Archives de Psychologie, 50,* 191–218.
Washburn, M. F. (1926). *The animal mind* (3rd ed.). New York: Macmillan.
Wesley, F. (1961). The number concept: A phylogenetic review. *Psychology Bulletin, 53,* 420–428.
Whiten, A., & Byrne, R. W. (1988). Tactical deception in primates. *The Behavioral and Brain Sciences, 11,* 169–183.

CHAPTER SEVEN

Do Animals Subitize?

Daniel J. Miller
Purdue University

Recent investigations in the animal learning literature have been concerned, to a varying degree, with the nature of the representational systems of animals, that is, how animals utilize internal copies or codes of physical stimuli in guiding their behavior (see, e.g., Hulse, Fowler, & Honig, 1978; Pearce, 1987; Roitblat, 1987; Roitblat, Bever, & Terrace, 1984). One area of interest currently receiving much attention, as evidenced by reports in this text, is the study of the numerical ability of animals. The fundamental question being addressed is whether the process underlying such ability is functionally equivalent to counting, as it is understood in the human literature. That is, does the performance of animals on numerical tasks reflect the process of applying internal number *tags* according to the principles specified by Gelman and Gallistel (1978) in their analysis of counting? Since the famous (or, rather, infamous) case of von Osten's supposedly mathematically inclined horse, Clever Hans, investigators of the numerical ability of animals have been highly sensitized to the possibility of animals employing (seemingly) low-level processes, in lieu of counting, to solve numerical discrimination problems (see Davis, chap. 5, and Rumbaugh, chap. 4, this vol.).

Subitizing is often viewed as being one such process. The term *subitize* first appeared in the psychological literature in a 1949 report by Kaufman, Lord, Reese, and Volkmann. Subitizing, derived from the classical Latin adjective *subitus,* "sudden," and the medieval Latin verb *subitare,* "to arrive suddenly," was proposed to account for the rapid, highly accurate, and confident numerosity judgment made to briefly presented visual stimulus fields containing six or fewer dots. The authors' intention was to distinguish this process from both *estimating,* believed to underlie the much less confident and less accurate judgment of arrays

149

containing more than six stimuli, and *counting,* the relatively slower process of enumerating one by one.

In a recent review of the animal numerical competence literature, Davis and Pérusse (1988a, 1988b) concluded that the numerical ability evidenced by animals should not be characterized as reflecting the operation of a "true" counting process but, rather, as reflecting the operation of other, cognitively less complex, processes. Specifically, the authors postulated that subitizing can account for much of the animal data.

The major concern of the present report is whether or not the majority of data from investigations of animal numerical competence can, in fact, be accounted for by the so-called subitizing process. What is currently known about subitizing, as it has been discussed in the human literature, is presented, followed by an examination into the extent to which subitizing has been investigated in the animal literature.

THE HUMAN LITERATURE

Procedures and Evidence

The study of numerosity judgment has employed the following procedures involving the simultaneous visual presentation of a collection of randomly arranged n objects, the subject responding verbally as to the number of objects presented. In the *threshold procedure,* stimulus exposure duration is brief (typically less than .5 s), the measure of interest being the percent correct (or error) as a function of n. In the *reaction time (RT) procedure,* an unlimited, subject-controlled exposure duration is used, the measure of interest being RT as a function of n. A third procedure is the synthesis of the previous two procedures—brief exposure duration and RT measured.

Use of the RT measure results in the following three possible patterns of responding in numerosity judgment tasks (see Klahr, 1973b, Fig. 1, p. 6): (a) *the continuous pattern*—as n increases, RT increases continuously; (b) *the discontinuous-flat pattern*—as n increases up to some number, R (the empirically determined upper limit of this range), RT is constant (i.e., slope = 0), and beyond R, RT increases as a function of n (i.e., slope > 0); and (c) *the discontinuous-increasing pattern*— as n increases up to R, RT increases but at a rate slower than that for $n > R$ (i.e., $n < R$ slope is less than $n > R$ slope). Plotting error data obtained from investigations using the threshold procedure can also result in the same patterns. Data fitting the discontinuous-flat and discontinuous-increasing patterns have led to the postulation of subitizing to account for the $n < R$ branch of the RT functions. From 1941 to 1950, three influential papers appeared in the literature, each reporting the need for two distinct processes for the discrimination of numerosity.[1] Taves (1941), Kaufman, Lord, Reese, and Volkmann (1949), and Jensen, Reese, and Reese (1950) reported discontinuities in RT, error, and/or confidence functions (confidence of each judgment was often assessed in the early reports) at about six, the data fitting the discontinuous-increasing pattern. Discontinuity may also be present in the functions reported by Saltzman and Garner (1948), although the authors interpreted their data as fitting the continuous pattern.

Discontinuity of data, however, does not necessarily imply the existence of two *qualitatively different* underlying *processes*. The operation of a single process, for example, counting, could also result in discontinuity of data. Counting the elements of a visual array requires the ability to distinguish already counted items from to-be-counted items—referred to as partitioning (see Gelman & Gallistel, 1978). The point of discontinuity may reflect the point at which partitioning begins to break down, resulting in the relatively sharp increase in RT and errors often reported. Thus, discontinuity alone is not sufficient for demanding the postulation of the operation of more than one process—more specific information is required. Such information is the particular value of the slope reported for the lower branch of the discontinuous RT function. If the judgment of arrays containing up to a certain number of elements is too rapid to be accounted for by the sequential application of counting tags to each of the elements, then postulation of the operation of a distinct process of numerosity judgment is required.

Klahr (1973b) provided such an analysis (see Allport, 1975, for a vehement criticism of Klahr's analysis). In Experiment 1, Klahr measured the time taken for subjects to verbally report the number of randomly arranged dots displayed on a computer monitor. The slope of the least squares linear regression line was 57 ms for 1–4 dots (increasing to 77 ms for 1–5 dots) and 300 ms for 5–20 dots. Klahr's (1973b) data were subjected to additional analysis by Klahr and Wallace (1976). Analyses of variance and trend analyses were performed for $n = 1-3$, $n = 1-4$, and so on, until a significant quadratic trend was obtained. The results of these analyses (performed over the data from Klahr, 1973b, Experiment 2) revealed the appearance of a significant quadratic trend at $n = 1-4$.

---

[1]The empirical question with which subitizing is concerned can be traced back to early philosophical inquiries. Philosophers debated the issue of whether or not the mind can consider more than a single object at any given time. From the late 1800s to the late 1920s, psychologists investigated the existence of a process for the immediate perception of numerosity. The so-called "span-of-apprehension" (or "attention") investigations attempted to determine the limit of such a process (for early reviews, see Graham, 1951; Woodworth, 1938; Woodworth & Schlosberg, 1954).

Thus, the subitizing range was stated to be 1 to 3, with the average subitizing rate determined by linear regression analysis to be 46 ms per dot. Other investigators have also determined the subitizing range to be about 1 to 4, operating at a rate of approximately 40 ms per item (see, e.g., Folk, Egeth, Kwak, 1988; Oyama, Kikuchi, & Ichihara, 1981). This rate is, arguably (cf. Gelman & Gallistel, 1978; Gallistel, 1990), too rapid to be accounted for by counting. In support of this, Klahr (1973b) cited the results of Olshavsky and Gregg (1970) purporting a 150 ms per item rate of implicitly reciting the alphabet from a specified starting point to a specified ending point. Landauer (1962) reported that subjects implicitly recited the numbers 1–10 at the rate of about 175 ms per number. Hence, another process has been inferred—subitizing.

Theoretical Accounts

The use of the term *subitizing* for describing the apparently distinctive process of rapidly discriminating small numbers of visually presented stimuli does not indicate an understanding of the underlying mechanism(s)—the process has simply been labeled. What is needed is a theoretical account of how this subitizing process operates. The status of current theorizing is reviewed, and additional relevant data are discussed.

A Production System/Template-Matching Account

Klahr (1973a, 1973b, 1984; Klahr & Wallace, 1973, 1976) took an information-processing approach to the theoretical analysis of subitizing (and quantification processes in general). According to Klahr (1973a, 1973b), subitizing is viewed as one of three subsystems of a larger *production system*. A production system consists of a number of individual rules, referred to as *productions*, taking the form of *condition–action* (if–then) *statements*. The stated action is carried out if the particular condition is satisfied by the symbolic information represented in short-term memory (STM; see, e.g., Newell & Simon, 1972). Specifically, subitizing is considered to be a rapid, serial comparison process in which a visually presented stimulus (e.g., array of dots) is "directly" matched to the proper quantitative symbol (representing numbers up to about 4) of the ordered "subitizing list" in STM (having been transferred from long-term memory, LTM). Each quantitative symbol in STM is associated with its respective number of "template pieces." Thus, the symbol "1" has one template piece, the symbol "2" has two template pieces, and so on. When each element of an array is paired with a template piece, the quantitative symbol is output as the response. According to Klahr (1973a, 1973b), the matching of the stimulus elements with the quantitative symbols of the subitizing list in STM is a *serial* self-terminating scanning process operating at about 40 ms per symbol, thus modeling the findings reported in the subitizing literature. The process of matching the elements of an array with the template pieces of a quantitative symbol is assumed to be *parallel* in nature. Thus, once a symbol is accessed, the time taken to check for a match is assumed to be independent of the number of template pieces comprising that symbol.

Klahr and Wallace (1976) later modified the model by extending the basic architecture of the system. At the front end of the process is a visual STM (VSTM) taking information from an iconic memory store (see, e.g., Sperling, 1960) and serially "unpacking" the undifferentiated group of elements of the array into individual elements. The information in VSTM is then compared with stored templates (i.e., productions in LTM), each associated with a specific quantitative symbol, until a match is obtained. The associated number is placed into semantic STM and the response is produced (also see Oyama, Kikuchi, & Ichihara, 1981).

Although this model can account for the rapid numerosity discrimination of arrays containing a small number of objects, it has been criticized on a number of grounds. Allport (1975) questioned the self-terminating nature of the model's memory search process, referring to data implicating an exhaustive memory search process. Chase (1978) emphasized what he believed to be the "loose connection between the theory and the data" (p. 69). Specifically, the problem in dealing with a production-system model consisting of numerous elementary condition-action rules is the objective determination of which of the elementary processes should be assigned temporal parameters and what their respective values should be. A second and perhaps more serious problem, according to Chase (1978), is the lack of a theoretical basis for a limit to the subitizing range. There is nothing stated in the operation of the control structure of the model that sets any such limit. Groups are formed in VSTM, but the model essentially says nothing with respect to how these groups are formed initially and, thus, nothing regarding their size limit (unless it were to be assumed that group size, and hence subitizing limit, is determined by the capacity limit and/or duration of either iconic memory or VSTM). Klahr (1973b) speculated that the size limit may reflect a limit of the template matching process if it were assumed that the actual templates are the unique configurations for arrays of n dots. There is one configuration for $n = 1$ (dot) and $n = 2$ (line), two for $n = 3$ (line, triangle), three for $n = 4$ (line, triangle, quadrilateral), and many for $n = 5$. Klahr (1984), however, argued against a pattern-recognition basis for subitizing.

Pattern Recognition Accounts

Subitizing has been viewed as the assignment of a number tag to a multi-element array on the basis of number; it has been assumed to be a *numerical process*. Contrary to this view, the following pattern-recognition accounts essentially consider subitizing to be a *nonnumerical process*. It is this approach that has received much attention in the animal literature.

Mandler and Shebo's Account.

Mandler and Shebo (1982) proposed that subitizing is a nonnumerical, perceptual process based on acquired canonical patterns (the notion that pattern cues may be used in the quantification of stimulus arrays has been noted by others: see, e.g., Messenger, 1903; Neisser, 1967; Warren, 1897; Woodworth & Schlosberg, 1954). That is, through experience "adults have developed canonical perceptions for arrays of 2 and 3—doublets (perceptually straight lines) and triplets (triangles)" (1982, p. 5). When random dot patterns are generated, the canonical pattern for 2 dots (straight line) always results, and the pattern for 3 dots (triangle) almost always results. For 4 dots, according to Mandler and Shebo (1982), the canonical pattern (square) does not frequently occur. They assume that the typical procedure of randomizing the dot patterns does not eliminate the availability of nonnumerical cues for the small, 2- and 3-dot arrays. Thus, for arrays of 1 to 3 dots, the adults can use the canonical pattern information to make rapid numerical judgments, resulting in the discontinuity of the RT function reported in the literature.

Mandler and Shebo (1982, Experiment 4) provided data supporting this view. In a within-subjects design, subjects judged the numerosity of both random and canonical arrays (the subjects were not informed of the presence of canonical patterns). For random arrays, a new pattern was randomly generated on each trial. For canonical arrays, a single pattern was used on each trial for each numerical value (for values 1–6, the patterns were those on the faces of a die; for patterns 7–10, see Mandler & Shebo, 1982, Fig. 9, p. 15). Importantly, it was reported that canonical-pattern RTs were flat for arrays of 1 to 5 elements (with essentially no errors). Additionally, there were no reliable differences in RTs between canonical and random patterns for arrays of 1–3, with shorter RTs and fewer errors on the former appearing at 4 and continuing through 10. Thus, the authors inferred that discrimination of random arrays of 1 to 3 elements is based on pattern recognition.

Although a flat slope for RT was reported for array sizes of 1 to 3 elements with canonical patterns, typically a slight increase in slope has been reported with random arrays. If the use of canonical patterns underlies the random-pattern data reported in the literature, the model needs to account for this increase. Though not directly addressed by Mandler and Shebo (1982), the increase in RT for 3 might be the result of longer times on those infrequent trials when "random generation will sometimes produce patterns clearly deviant from a canonical triangle or an obvious triple" (1982, p. 15). Why, then, is there an increase in RT from 1 to 2 in many reports using random arrays when presentation of 2 elements always results in a linear pattern? Possibly, the distance between the elements and/or the orientation of the linear patterns might affect RT. Regarding this issue, Davis and Pérusse (1988a, p. 604) stated that comparison against a pattern template may take longer for some particular patterns than for others, but, nevertheless, a pattern-recognition process would be operating.

Although this pattern-recognition view appears to account for the subitizing results quite well, the supporting data are rather limited and a number of issues have yet to be fully addressed. For instance, what is the nature of the perceptual pattern-recognition process? Apparently, according to Mandler and Shebo, "patterns of small numbers are familiar and responded to 'inferentially'; people 'know' that a triangular pattern represents 3 and respond accordingly" (1982, p. 19; see also, Wolters, van Kempen, & Wijhuizen, 1987). A question arises: Would the response "three" be rapidly given to any array in which the elements are arranged in a triangular pattern? That is, to what extent is the operation of the pattern-recognition process truly independent of the number of elements in the array? Also, how would this model account for rapid numerosity judgment of small linearly arranged arrays (with controls present for other nonnumerical cues, e.g., spacing)? Similarly, it is unclear how a pattern-recognition process could account for subitizing results obtained with relatively complex Shepard–Metzler block configurations as stimuli (Chase, 1978). Though pattern recognition may contribute to the rapid discrimination of visual arrays, more empirical and theoretical analysis is required before concluding that subitizing reflects the operation of a truly nonnumerical, pattern-recognition process.

Von Glasersfeld's Account. Von Glasersfeld (1982; also see von Glasersfeld, chap. 11, this vol.) also viewed subitizing as a perceptual process, rather than a numerical process, based on figural patterns. Before children are able to count, they are said to be able to associate a particular verbal numerical label with a perceptual figural pattern resulting from the recurrent arrangement of a number of objects (e.g., fingers of the hands, dominoes, playing cards). Thus, subitizing is essentially a *semantic* labeling process. The perception of figural patterns is based on visual "scan-paths." It is suggested that characteristic scan-paths exist for arrays of 2 (straight), 3 (triangular), 4 (quadrilateral), and certain arrays of 5 elements utilized in the determination of numerosity of multielement arrays. Regarding the discrimination of linearly arranged arrays for all of which there would be a straight scan-path, von Glasersfeld (1982) proposed:

There must be some inherent difference in the process of perceiving two as opposed to three dots. That difference, I suggest, arises from the capability of the nervous system to differentiate between the temporal patterns constituted by a succession of two perceptual acts as opposed to one or three. . . . In other words, I am positing a neural mechanism that distinguishes a dual incidence of a given event from a single or a treble incidence. (p. 202)

Von Glasersfeld (1982) used the word *rhythm*, as it has been used in music perception (e.g., Fraisse, 1982), to refer to this process of temporal-pattern perception. He cited evidence that children can recognize and replicate simple rhythms prior to their having conventional, notational representations of them, and can graphically represent auditory patterns (e.g., Bamberger, 1980, 1982).

Von Glasersfeld's (1982) account is speculative, however, lacking direct empirical support. There are no data uniquely implicating the operation of a perceptual scan-path process for rapid determination of numerosity of arrays consisting of relatively small numbers of randomly arranged discrete elements. The same can be said regarding the proposed neural mechanism for temporal-pattern perception of linear arrays not affording discrimination on the basis of distinctive scan-paths. Why, then, postulate the operation of perceptual and neural mechanisms for detecting differences among perceptual scan-paths and temporal patterns, respectively, instead of postulating a single mechanism for registering *number* of discrete elements and, thus, quantifying arrays on a numerical rather than pattern basis (see Thompson, Mayres, Robertson, & Patterson, 1970)? It suggests an unfounded bias against postulating mechanisms of numerosity detection.

A Specialized Rapid Counting View

Gelman and Gallistel (1978) proposed that the so-called subitizing data can be accounted for by the postulation of a specialized rapid *counting* strategy (see also, Gallistel, 1988, 1990). The RT data reported in the literature are assumed to be the product of a rapid, subvocal, and apparently unconscious serial enumeration process, rather than the product of a rapid, nonserial perceptual process. Thus, as is the case in most reports, RT increases even from 1 to 2 and 2 to 3. Development of this rapid counting strategy begins when children have progressed from overt to covert counting in determining the numerosity of small arrays. Gelman and Tucker (1975) reported that 3-year-old children counted aloud when estimating arrays of 2 and 3 elements, whereas 4- and 5-year-olds counted aloud only when estimating larger 4- and 5-element arrays. According to Gelman and Gallistel (1978), the transition to covert counting may allow for the "routinization" of counting.

It is clear that a specialized, rapid, but nevertheless serial, counting process could account for the subitizing data. The upper limit of subitizing, as defined by RT-slope discontinuity, is accounted for by postulating a limit to the range of the specialized counting strategy, beyond which formal counting would operate. Unfortunately, Gelman and Gallistel (1978) did not specify precisely how this routinized counting strategy operates and specifically differs from counting proper. The authors only speculate that the limit of such a process may "have something to do with the rhythmic organization of action. It may be no accident that most of our musical rhythms are based on groups of no more than three or four beats" (1978, p. 224). In sum, the existence of a specialized counting process awaits empirical determination.

Gelman and Gallistel (1978) noted, however, that nothing in their account precludes the use of perceptual information in determining the numerosity of arrays (e.g., using multiplication to quickly arrive at a judgment of "16" for a 4 × 4 array; see, e.g., Beckwith & Restle, 1966). Importantly, they stressed that the ability to use such information requires an understanding of perceptual relations and, therefore, should be considered a *skill* developed to aid in counting, rather than a low-level process. Thus, they concluded that even if the hypothesis of a rapid counting process is not accepted, subitizing should not be defined as a low-level primitive process.

It is clear that our understanding of subitizing is incomplete. More empirical investigation is needed, accompanied by greater specificity in the theorizing of the underlying mechanism(s) of subitizing.

Subitizing in Nonadults: The Developmental Relationship Between Subitizing and Counting

Since the work of Piaget (1941/1965), an extensive body of literature has been amassed on the development of numerical abilities (e.g., Fuson, 1988; Gelman & Gallistel, 1978). Of concern here is the developmental relationship between subitizing and counting; does subitizing serve as a developmental precursor to counting, or vice versa?

According to Klahr (see, e.g., Klahr & Wallace, 1976) and von Glasersfeld (1982), subitizing precedes counting. In Klahr's (Klahr & Wallace, 1976) production system model, subitizing develops and operates prior to "the acquisition of socially transmitted technology" (p. 67), such as conventional verbal tags said to be required for counting. Subitizing is also said to be the process that first produces consistent outcomes in response to numerical arrays. Von Glasersfeld (1982) viewed subitizing as a semantic labeling process operating prior to the acquisition of any "conception of number as units composed of units" (p. 207). Mandler and Shebo (1982), on the other hand, posed that subitizing develops following the acquisition of counting. Subitizing is viewed as a perceptual, pattern-recognition process based on canonical patterns acquired subsequent to counting. According to Gelman and Gallistel (1978), the so-called subitizing data reflect the operation of a rapid, routinized counting strategy that develops as children progress from overt to covert counting.

If subitizing developmentally precedes counting, evidence for subitizing should exist in the very young. It appears, however, that little, if any, such evidence exists. Chi and Klahr (1975), using the procedure employed by Klahr (1973b), reported discontinuity in the slope of the RT function at $n = 4$ for 5- to $6\frac{1}{2}$-year-old children. As the authors noted, however, the obtained subitizing rate (195 ms per dot) is comparable to the children's numerical-recitation rate; the subitizing rate for adults is about four times faster than their numerical-recitation rate. Thus, the children might have counted the small, 2- and 3-dot arrays. Svenson and Sjöberg (1978, 1983) also examined the range and speed of subitizing in children (7- to 15-year-olds). However, the RT data are difficult to interpret due to various methodological problems (e.g., spacing was not varied between the linearly arranged dot stimuli, thus confounding number of dots with array length).

Subitizing is often invoked in instances where children determine the numerosity of briefly presented arrays containing relatively small numbers of elements without engaging in overt counting (e.g., Fuson, 1988; Young & McPherson, 1976). There is an implicit assumption that, in the absence of overt counting and pointing at or touching the elements as they are being enumerated (referred to as *motor tagging*), a distinct subitizing process underlies numerosity judgment of briefly presented arrays. More information is required, however, before reaching such a conclusion; it is possible that covert counting underlies performance. Thus, the "subitizing" limit might represent the number of briefly presented objects that can be either counted directly, that is, while they are physically present, or held in memory and counted. Absence of overt counting is not necessarily evidence of the operation of a nonnumerical, perceptual process. Again, in the literature on subitizing in adults, the operation of a process distinct from counting has been invoked when numerosity judgment has been believed to be too rapid to be accounted for by counting. Subitizing should not be called on to characterize numerosity discrimination in children when RT data are unavailable.

In addition to the current lack of substantial evidence for subitizing in children, Gelman and co-workers provided further evidence for the counting-to-subitizing relationship. Gelman (1972) reported that children tended to count aloud when presented a small array of items that had undergone some type of change from an earlier presentation (e.g., spacing or number was altered). Recall that Gelman and Tucker (1975) reported that 3-year-old children counted aloud when estimating arrays of 2 and 3 elements, whereas 4- and 5-year-old children counted aloud only when estimating larger 4- and 5-element arrays. Also, for the youngest children, longer stimulus exposure duration led to increased accuracy for the smaller and larger arrays, whereas, for the older children, the benefit of increased exposure duration was found only for larger arrays. Finally, Gelman and Gallistel (1978) observed and documented the apparent salience of counting in their many hours of testing the numerical ability of young children on a variety of experimental tasks. Included are the reports of spontaneous counting, children asking if they can count when performing certain tasks, and children's tendencies to count under conditions affording counting (e.g., relatively long stimulus exposure duration), complaining that they cannot count under conditions tending to prohibit counting (e.g., relatively short exposure duration; see also, Fuson, 1988, p. 219; Silverman & Rose, 1980).

Recently, the ability of infants to discriminate arrays on the basis of their numerosity has been investigated. Using the habituation–dishabituation procedure (see, e.g., Friedman, 1972), infants are first habituated to an array of a given number of elements, then presented with an array consisting of either the same or a different number of elements. Dishabituation to arrays of a different number of elements, but not to arrays of the same number of elements, is taken as evidence of numerosity discrimination (provided there is no nonnumerical basis for discrimination, such as brightness, area, and density). Starkey and Cooper (1980) reported discrimination between 2 and 3, but not between 4 and 6 (16- to 30-week-old infants). Antell and Keating (1983) reported similar findings using neonates (mean age of 53 hours old). Strauss and Curtis (1981) reported discrimination between 2 and 3 and 3 and 4, but not between 4 and 5 (10- to 12-month-old infants).

Some investigators concluded that such discrimination suggests the operation of a perceptual subitizing process; the range of discrimination is comparable to the subitizing range in adults, and the infants, obviously, cannot verbally count (see, e.g., Starkey & Cooper, 1980). Is it not possible, however, that infants possess some form of nonverbal counting (as is currently being debated in the study of animal numerical ability; see, however, Strauss & Curtis, 1984)? That is, infants might have a limited set of nonverbal, internal numerical representations or tags (what Gelman & Gallistel, 1978, referred to as "numerons" and Koehler, 1950, referred to as "unnamed numbers") they can use for discriminating arrays containing relatively small numbers of elements (see Starkey, Spelke,

& Gelman, 1980, cited in Klahr, 1984). The limit of their discrimination ability could be due to either a limited ability in generating distinct number tags or in applying them appropriately (i.e., keeping tagged and non-tagged elements separated). This view might be difficult to accept in light of data indicating rapid numerosity detection in infants (Cooper, 1984). Such preliminary evidence, however, only begins to address the possibility of the presence of a rapid, subitizinglike process of numerical detection in young infants.

THE ANIMAL LITERATURE

Though an adequate understanding of subitizing is lacking, Davis and Pérusse (1988b) invoked subitizing to account for much of the data on numerical competence in animals. Interestingly, in an earlier, highly influential review of animal "counting behavior," Davis and Memmott (1982) stated that to do so is "misdirected in the case of infrahuman evidence" (p. 549). The results of the following investigations are representative of the data for which Davis and Pérusse (1988a, 1988b) gave a subitizing interpretation. The reports are briefly reviewed, for the intent here is not to determine whether the data from these studies implicate the operation of a "true" counting process, but to evaluate the proposed subitizing account of the data.

Matsuzawa (1985) examined a chimpanzee's ability to correctly assign the Arabic numeral corresponding to the number of objects displayed (e.g., pencils, spoons). On some trials, the objects (1 to 6) were presented in a linear arrangement, with position and spatial density varied. On other trials, the objects were held in the experimenter's hand with orientation and angular separation of the items varied. Shifts to novel objects were carried out. It was reported that performance on original and novel objects was well above chance.

Thomas, Fowlkes, and Vickery (1980) trained two squirrel monkeys on a two-choice discrimination task. Pairs of stimuli (filled circles on cards) were presented with reinforcement delivery contingent on the choice of the card with the lesser number of circles. Controls for area, brightness, and pattern cues were carried out. Training proceeded in the order 2 versus 7 (2:7), 2:6 ... 2:3, 3:7, 3:6, and so on up to 8:9 (or until failure to meet criterion of 75% correct in 500 trials). It was reported that both monkeys met criterion up through 7:8, with one of the monkeys also meeting criterion on 8:9.

In a well-controlled set of experiments, Pepperberg (1987) reported on the ability of an African Grey Parrot, Alex, to correctly apply learned vocal English labels for the numerals 1–6 to arrays of various objects (e.g., corks, keys). The objects were presented either by hand (with orientation and angular separation varied) or resting on a tray (either linearly arranged, with spacing varied, or randomly arranged). Additionally, heterogenous trials were run such that Alex was required to give the number of only one of the two types of objects presented in a "mixed" array (e.g., an array of 5 objects, 3 corks and 2 keys, intermixed). The types of objects used, along with the various modes of presentation and the use of transfer trials involving novel objects, all served to reduce, if not eliminate, the presence of area, brightness, and pattern cues. Pepperberg (1987) reported that Alex performed at levels reliably above chance throughout training.

Boysen and Berntson (1989) presented some remarkable data on the ability of a chimpanzee, Sheba, to sum arrays of food items. On each trial, Sheba visited three food sites in the lab. One or two of the sites contained oranges. The total number of oranges on a given trial varied from 1 to 4. After visiting the three sites, Sheba was tested on her ability to choose the Arabic numeral (Figures 1–4 printed on separate cards) that corresponded to the total number of oranges viewed. Subsequently, employing the same general procedure, the oranges were replaced with the number cards (0–4) at the three sites. Sheba's performance on both tasks was reported to be well above chance.

As stated, Davis and Pérusse (1988b) claimed that the numerical discrimination evidenced in the previous reports can be accounted for by subitizing, in that each involves the discrimination of a relatively small number of simultaneously presented items. Let us accept, for the moment, that the animals in the previous studies have a subitizing process available to them. In the view of Davis and Pérusse (1988a, 1988b), this process would be pattern-recognition based. They concurred with the view of subitizing expressed by von Glasersfeld (1982), and defined subitizing as the "rapid assignment of a numerical tag to small quantities of items (typically less than six). Subitizing is a form of perceptual shorthand based on pattern recognition and labeling rather than formal enumeration" (1988b, p. 562). Though a limit to subitizing was not discussed by von Glasersfeld (1982), the pattern-recognition view of Mandler and Shebo (1982) set the limit of such a process at 3. Again, it is generally thought that for random arrays of more than 4 objects, there are simply too many possible resulting configurations, making discrimination based on pattern difficult, if not impossible. Recall that a subitizing limit of about 4 has been determined by the results of investigations using the RT procedure. How then, based on the view that Davis and Pérusse (1988a, 1988b) appear to accept, can they logically conclude that the aforementioned reports are subject to a subitizing interpretation? Matsuzawa (1985) and Pepperberg (1987) reported above-chance performance with arrays up through 6. Thomas and colleagues (1980) reported accurate discrimination between arrays of 7 and 8 items, and between 8 and 9 items by one of their monkeys! These array sizes appear to lie beyond the limit set by a pattern-recognition process, as it has been discussed in the human literature. Davis and Pérusse (1988b) accounted for this apparent discrepancy between the upper limit of human and nonhuman subitizing by referring to R. K. Thomas's (personal communication to H. Davis, 1987) proposal that the unlimited stimulus exposure duration employed in the animal investigations afforded discrimination between relatively large arrays. This explanation is insufficient; it does not address the specific nature of the distinctive pattern information present in *random* arrays of 6, 7, and 8 items, for example, that even with unlimited processing time, would afford discrimination. Also, consider the data reported by Boysen and Berntson (1989). The to-be-counted objects were not presented simultaneously, but rather in "groups" separated by a time interval. Recall that Arabic numerals were substituted for the discrete objects. How would a pattern-recognition process account for these data? Davis and Pérusse (1988a) subsequently concluded, in response to Boysen (1988), that subitizing could not account for the data reported by Boysen and Berntson (1989).

If a pattern-recognition process is to be proposed as the mechanism by which animals are able to make numerical discriminations, then there should be some independent empirical evidence of pattern recognition in animals. It is apparent that animals can learn to discriminate among stimuli on the basis of their shape. Lashley (1938) demonstrated that rats can discriminate among various geometric shapes (e.g., circles and triangles). Pigeons and monkeys perform matching-to-sample tasks using geometric figures as stimuli (e.g., D'Amato, Solmon, & Colombo, 1985). There is the recent work on pattern or form recognition in birds (e.g., Blough, 1984; Herrnstein, 1984; Steirn, 1985). Also, there is the body of literature on language learning in primates, where lexigrams of the artificial Yerkish language were used and obviously discriminated (e.g., Rumbaugh, 1977). These data may not be directly relevant, however, to the discussion of pattern recognition in situations involving arrays of discrete elements. It is one thing to conclude that animals can discriminate on the basis of the geometric shape of a solid figure; it is quite another thing entirely to conclude that animals are abstracting and utilizing pattern information in discriminating among arrays of discrete elements, particularly when those elements are randomly or linearly arranged. Recall that Mandler and Shebo (1982) viewed subitizing as a pattern-recognition process acquired by adults following the acquisition of counting. Davis and Pérusse (1988a, 1988b) accepted the pattern-recognition account, yet did not seriously consider the possibility that such a process might not be "simpler" than counting (see Gallistel, 1988, 1990; Macphail, 1988).

Even if it were to be determined that animals could discriminate such arrays on the basis of their pattern arrangement, it would not necessarily preclude the possibility that the arrays could also be discriminated on the basis of their number. Although Davis and Pérusse's view (1988a, 1988b) implies that pattern cues will always be used in lieu of, that is, *overshadow,* numerical cues, no data exist to support such a position. It would have to be determined by systematic investigation the conditions under which animals (and which particular animals) utilize the pattern information, numerical information, or both. As Davis and Memmott (1982) astutely warned—in reaction to Morgan's canon—in their review, "do not exclude on a priori grounds the possibility of control by seemingly more complex stimulus dimensions" (p. 549).

To this point, the animal data have been evaluated in terms of a pattern-recognition view of subitizing. The more general concern, without considering any particular view of subitizing, is the nature of the data compelling one to conclude that a perceptual, subitizing process is operating. In the human literature, subitizing has been proposed to account for the rapid and accurate assignment of numerical tags to visually presented arrays of a relatively small number of elements. Again, it is the view that the RTs for small arrays, up to some limit, preclude the operation of formal counting (recall, however, Gelman & Gallistel's, 1978, postulation of a specialized rapid-counting process). In the animal numerical literature, however, RT data have not been collected. What is the basis, then, for postulating the operation of a subitizing process to account for the animal data?

It appears that Davis and Pérusse's (1988a, 1988b) criterion for inferring subitizing is based simply on the paradigm used: If a small number of visual elements are simultaneously presented, then subitizing is most likely the basis for the animal's discrimination performance (although "small" can be quite large, e.g., Thomas et al., 1980). Davis and Pérusse (1988a, 1988b) essentially abandoned the measures of speed and confidence in their analysis of nonhuman subitizing (see Thomas, 1988). Their view that the confidence measure is not really appropriate with nonhuman subjects is reasonable. Regarding the speed measure, the authors stated that the "[speed] criterion was crucial only when a subject's ability to count posed a serious alternative explanation. To date, this does not seem to be a problem with most nonhuman subjects" (1988a, p. 603). Hence, what we are left with as evidence for a perceptual subitizing process is the accurate discrimination of arrays consisting of a relatively small number of elements. A counting process, however, could also account for such discrimination, although Davis and Pérusse (1988a, 1988b) concluded that their review of the literature revealed no evidence of true counting in animals. Clearly, before subitizing is invoked as the process underlying animal numerical behavior, an experimental methodology that empirically demonstrates the existence of an accurate (and possibly rapid) perceptual, nonnumerical process of numerosity discrimination in animals—operating under the conditions and at the limits delineated in the previous reports—is needed. Such evidence is currently lacking. Thus, it should not be concluded that animals have available to them a subitizing process, simply because there is some evidence that humans possess such a process. The same experimental rigor demanded for demonstrating counting in animals is similarly demanded for demonstrating subitizing in animals.

Subitizing of Sequentially Presented Items

The possibility that animals may use available temporal-rhythm cues in solving numerical tasks involving the sequential presentations of items has been recognized for many years (e.g., Salmon, 1943; Thorpe, 1956). Davis and Pérusse (1988a, 1988b) proposed that a series of events presented in a regular temporal order can be learned on the basis of temporal regularity. They accepted von Glasersfeld's (1982) extension of the subitizing process to situations involving *sequential* presentation of items (i.e., only one of the to-be-enumerated items is physically present at any given time), using the term *rhythm* to refer to such "sequential subitizing." In their view, rhythm is a temporally based, pattern-matching process. Unlike the subitizing of simultaneously presented arrays, however, the use of rhythm may not be limited to small quantities, although a limit most likely exists (unspecified by the authors). Also, according to this view, if the time interval between events is relatively long (also unspecified), then a pattern will not be detected, thus prohibiting discrimination on the basis of rhythm.

Davis and Pérusse (1988b) reviewed a number of investigations of animal numerical ability involving successively presented events, for which they concluded that the use of rhythm may have been the basis for performance. Rather than review each of these studies, I examine one such report to illustrate their rhythm interpretation.

In a series of experiments, Capaldi and Miller (1988b; see also Burns & Sanders, 1987) examined the ability of the rat to count successively presented reward events. On successive runs down a straight runway, rats received either food reinforcement (R) or nonreinforcement (N). In Experiment 1, the rats received training on two series, RRN and NRRN, each presented three times daily in an irregular order. There was a short (10–15 s) interval between the trials within a series (the intertrial interval or ITI), with a longer (10–15 min) interval elapsing between series. With training, the rats learned to run slowly in anticipation of the terminal nonreinforced trial of both series, while running rapidly on all other trials. Capaldi and Miller (1988b) concluded that the rats were using number of R trials to predict the terminal-N trials.

Davis and Pérusse (1988b), however, concluded that the rat's performance was most likely based on the use of rhythm, because there was a relatively constant and short time interval between trials of the series. How can this be? In the example given, the presumed rhythmic pattern predicting the terminal-N trial in the RRN series (Trial 3) would presumably predict reinforcement in the NRRN series (Trial 3). It is possible that the rat learned two rhythmic patterns—one for the series initiating with R, another for the series initiating with N. Or that the initial-N trial of the NRRN series was somehow disregarded by the rat and not included in the determination of the rhythmic pattern. This latter suggestion modifies the rhythm interpretation to the extent that it is no longer solely the encoding of temporal regularity of events.

Other data in the report, however, appear to further render a rhythm account untenable (see Burns, 1988; Capaldi & Miller, 1988a). Performance was reported to be unaffected by procedural shifts involving increases in goal box confinement duration on R trials, increases in ITI, and shifts to novel series in which the rhythmic pattern predicting terminal-N trials was changed (yet number of preceding R trials remained unchanged). Such manipulations should have disrupted performance if it were based on the use of rhythmic cues. Davis and Pérusse (1988a, 1988b) viewed these manipulations as insufficient, having indicated that the use of varied and/or relatively long ITIs is necessary to rule out the rhythm interpretation.

Capaldi and co-workers (unpublished raw data; see Capaldi, chap. 9, this vol.) recently examined the effect of variable ITIs on the rat's performance on RRN and NRRN series. It was found that varying the ITIs within a series had no effect on the anticipation of the terminal-N trials. Also, a shift from constant ITIs to varied ITIs did not disrupt performance. If rhythm were controlling anticipation of the terminal-N trials under the constant-ITI condition, then removal of the rhythm cues in the shift to varied ITIs should have eliminated correct anticipation.

It is important to note that even when rhythm cues are available, they may not always be used in lieu of using other available cues, contrary to Davis and Pérusse's (1988a) unsupported belief that "rhythmic pattern represents the simplest and most robust source of discriminative information" (p. 605). Again, in the absence of empirical data, it should not be assumed that nonnumerical cues will always overshadow numerical cues.

CONCLUSIONS

Subitizing has been proposed as a label for the rapid and highly accurate numerosity judgment of stimulus fields containing relatively small numbers of elements. A review of the human literature reveals that more empirical and theoretical work is needed before the underlying mechanism(s) of subitizing is (are) fully understood. A review of the animal literature reveals that the position that much of the animal numerical data can be accounted for by a perceptual subitizing process, as it is currently understood in the human literature, is untenable. There are currently no data in the animal literature comparable to those in the human literature that have led to the postulation of subitizing. Thus, it has yet to be determined whether or not animals subitize. Additionally, various aspects of the animal investigations, such as the specific controls employed and the empirically determined range of numerical discrimination, appear to rule out a subitizing interpretation. Modifying the definition of subitizing by eliminating its traditional defining characteristics (speed and range of operation) to account for the animal data is unacceptable. If the available data are to be accounted for by a yet unspecified nonnumerical process, a term other than subitizing should be used to refer to such a process: The label *subitizing* should not be incautiously invoked to characterize animal numerical ability.

REFERENCES

Allport, D. A. (1975). The state of cognitive psychology, *Quarterly Journal of Experimental Psychology, 27,* 141–152.

Antell, S. E., & Keating, D. P. (1983). Perception of numerical invariance in neonates. *Child Development, 54,* 695–701.

Bamberger, J. (1980). Cognitive structuring in the apprehension and description of simple rhythms. *Archives de Psychologie, 48,* 171–197.

Bamberger, J. (1982). Revisiting children's drawings of simple rhythms: A function for reflection-in-action. In S. Strauss (Ed.), *U-shaped behavior growth* (pp. 191–226). New York: Academic Press.

Beckwith, M., & Restle, F. (1966). Process of enumeration. *Psychological Review, 73,* 437–444.

Blough, D. S. (1984). Form recognition in pigeons. In H. L. Roitblat, T. G. Bever, & H. S. Terrace (Eds.), *Animal cognition* (pp. 277–289). Hillsdale, NJ: Lawrence Erlbaum Associates.

Boysen, S. T. (1988). Kanting processes in the chimpanzee: What (and who) really counts? *Behavioral and Brain Sciences, 11,* 580.

Boysen, S. T., & Berntson, G. G. (1989). Numerical competence in a chimpanzee *(Pan troglodytes). Journal of Comparative Psychology, 103,* 23–31.

Burns, R. A. (1988). Subitizing and rhythm in serial numerical investigations with animals. *Behavioral and Brain Sciences, 11,* 581–582.

Burns, R. A., & Sanders, R. E. (1987). Concurrent counting of two and three events in a serial anticipation paradigm. *Bulletin of the Psychonomic Society, 25,* 479–481.
Capaldi, E. J. (1992). [A number discrimination with rhythmic cues present or absent.] Unpublished raw data.
Capaldi, E. J., & Miller, D. J. (1988a). A different view of numerical processes in animals. *Behavioral and Brain Sciences, 11,* 582–583.
Capaldi, E. J., & Miller, D. J. (1988b). Counting in rats: Its functional significance and the independent cognitive processes that constitute it. *Journal of Experimental Psychology: Animal Behavior Processes, 14,* 3–17.
Chase, W. G. (1978). Elementary information processes. In W. K. Estes (Ed.), *Handbook of learning and cognitive processes* (pp. 19–90). Hillsdale, NJ: Lawrence Erlbaum Associates.
Chi, M. T. H., & Klahr, D. (1975). Span and rate of apprehension in children and adults. *Journal of Experimental Child Psychology, 19,* 434–439.
Cooper, R. G., Jr. (1984). Early number development: Discovering number space and addition and subtraction. In C. Sophian (Ed.), *Origins of cognitive skills* (pp. 157–192). Hillsdale, NJ: Lawrence Erlbaum Associates.
D'Amato, M. R., Salmon, D. P., & Colombo, M. (1985). Extents and limits of the matching concept in monkeys *(Cebus apella). Journal of Experimental Psychology: Animal Behavior Processes, 11,* 35–51.
Davis, H., & Memmott, J. (1982). Counting behavior in animals: A critical evaluation. *Psychological Bulletin, 92,* 547–571.
Davis, H., & Pérusse, R. (1988a). Numerical competence: From backwater to mainstream of comparative psychology. *Behavioral and Brain Sciences, 11,* 602–615.
Davis, H., & Pérusse, R. (1988b). Numerical competence in animals: Definitional issues, current evidence, and a new research agenda. *Behavioral and Brain Sciences, 11,* 561–579.
Folk, C. L., Egeth, H. E., & Kwak, H. (1988). Subitizing: Direct apprehension or serial processing? *Perception & Psychophysics, 44,* 313–320.
Fraisse, P. (1982). Rhythm and tempo. In D. Deutsch (Ed.), *The psychology of music* (pp. 149–180). New York: Academic Press.
Friedman, S. (1972). Habituation and recovery of visual response in the alert human newborn. *Journal of Experimental Child Psychology, 13,* 339–349.
Fuson, K. C. (1988). *Children's counting and concepts of number* (Springer series in cognitive development). New York: Springer-Verlag.
Gallistel, C. R. (1988). Counting versus subitizing versus the sense of number. *Behavioral and Brain Sciences, 11,* 585–586.
Gallistel, C. R. (1990). *The organization of learning.* Cambridge, MA: MIT Press.
Gelman, R. (1972). The nature and development of early number concepts. In H. W. Reese (Ed.), *Advances in child development and behavior* (Vol. 7, pp. 115–167). New York: Academic Press.
Gelman, R., & Gallistel, C. R. (1978). *The child's understanding of number.* Cambridge, MA: Harvard University Press.
Gelman, R., & Tucker, M. F. (1975). Further investigations of the young child's conception of number. *Child Development, 46,* 167–175.
Graham, C. H. (1951). Visual perception. In S. S. Stevens (Ed.), *Handbook of experimental psychology* (pp. 868–920). New York: Wiley.
Herrnstein, R. J. (1984). Objects, categories, and discriminative stimuli. In H. L. Roitblat, T. G. Bever, & H. S. Terrace (Eds.), *Animal cognition* (pp. 233–261). Hillsdale, NJ: Lawrence Erlbaum Associates.
Hulse, S. H., Fowler, H., & Honig, W. K. (Eds.). (1978). *Cognitive processes in animal behavior.* Hillsdale, NJ: Lawrence Erlbaum Associates.

Jenson, E. M., Reese, E. P., & Reese, T. W. (1950). The subitizing and counting of visually presented fields of dots. *The Journal of Psychology, 30,* 363–392.
Kaufman, E. C., Lord, M. W., Reese, T. W., & Volkmann, J. (1949). The discrimination of visual number. *American Journal of Psychology, 62,* 498–525.
Klahr, D. (1973a). A production system for counting, subitizing and adding. In W. G. Chase (Ed.), *Visual information processing* (pp. 527–546). New York: Academic Press.
Klahr, D. (1973b). Quantification processes. In W. G. Chase (Ed.), *Visual information processing* (pp. 3–34). New York: Academic Press.
Klahr, D. (1984). Transition processes in quantitative development. In R. J. Sternberg (Ed.), *Mechanisms of cognitive development* (pp. 101–139). New York: W. H. Freeman.
Klahr, D. R., & Wallace, J. G. (1973). The role of quantification operators in the development of conservation. *Cognitive Psychology, 4,* 301–327.
Klahr, D. R., & Wallace, J. G. (1976). *Cognitive development: An information processing view.* Hillsdale, NJ: Lawrence Erlbaum Associates.
Koehler, O. (1950). The ability of birds to "count." *Bulletin of Animal Behaviour, 9,* 41–45.
Landauer, T. K. (1962). Rate of implicit speech. *Perceptual and Motor Skills, 15,* 646.
Lashley, K. S. (1938). The mechanism of vision: XV. Preliminary studies of the rat's capacity for detail vision. *Journal of General Psychology, 18,* 123–193.
Macphail, E. M. (1988). You can't succeed without really counting. *Behavioral and Brain Sciences, 11,* 592–593.
Mandler, G., & Shebo, B. J. (1982). Subitizing: An analysis of its component processes. *Journal of Experimental Psychology: General, 11,* 1–22.
Matsuzawa, T. (1985). Use of numbers by a chimpanzee. *Nature, 315,* 57–59.
Messenger, J. F. (1903). The perception of number. *Psychological Monographs, 5,* No. 22.
Neisser, U. (1967). *Cognitive psychology.* New York: Appleton-Century-Crofts.
Newell, A., & Simon, H. A. (1972). *Human problem solving.* Englewood Cliffs, NJ: Prentice-Hall.
Olshavsky, R. W., & Gregg, L. W. (1970). Information processing rates and task complexity. *Journal of Experimental Psychology, 83,* 131–135.
Oyama, T., Kikuchi, T., & Ichihara, S. (1981). Span of attention, backward masking, and reaction time. *Perception & Psychophysics, 29,* 106–112.
Pearce, J. M. (1987). *Introduction to animal cognition.* Hillsdale, NJ: Lawrence Erlbaum Associates.
Pepperberg, I. M. (1987). Evidence for conceptual quantitative abilities in the African Grey Parrot: Labeling of cardinal pets. *Ethology, 75,* 37–61.
Piaget, J. (1965). *The child's conception of number.* New York: W. W. Norton. (Original work published 1941)
Roitblat, H. L. (1987). *Introduction to comparative cognition.* New York: W. H. Freeman.
Roitblat, H. L., Terrace, H. S., & Bever T. G. (Eds.). (1984). *Animal cognition.* Hillsdale, NJ: Lawrence Erlbaum Associates.
Rumbaugh, D. M. (1977). *Language learning by a chimpanzee: The Lana project.* New York: Academic Press.
Salman, D. H. (1943). Note on the number conception in animal psychology. *British Journal of Psychology, 33,* 209–219.
Saltzman, I. J., & Garner, W. R. (1948). Reaction-time as a measure of span of attention. *The Journal of Psychology, 25,* 227–241.
Silverman, I. W., & Rose, A. P. (1980). Subitizing and counting skills in 3-year-olds. *Developmental Psychology, 16,* 539–540.
Sperling, G. (1960). The information available in brief visual presentations. *Psychological Monographs, 74,* No. 498.
Starkey, P., & Cooper, R. J. (1980). Perception of numbers by human infants. *Science, 210,* 1033–1035.

Steirn, J. (1985). *Concept formation in Japanese Quail* (Coturnix coturnix japonica). Unpublished doctoral dissertation, University of Georgia.
Strauss, M. S., & Curtis, L. E. (1981). Infant perception of numerosity. *Child Development, 52*, 1146–1152.
Strauss, M. S., & Curtis, L. E. (1984). Development of numerical concepts in infancy. In C. Sophian (Ed.), *Origins of cognitive skills* (pp. 131–155). Hillsdale, NJ: Lawrence Erlbaum Associates.
Svenson, O., & Sjöberg, K. (1978). Subitizing and counting processes in young children. *Scandinavian Journal of Psychology, 19*, 247–250.
Svenson, O., & Sjöberg, K. (1983). Speeds of subitizing and counting processes in different age groups. *Journal of Genetic Psychology, 142*, 203–211.
Taves, E. H. (1941). Two mechanisms for the perception of visual numerousness. *Archives of Psychology, 37*, No. 265.
Thomas, R. K. (1988). To honor Davis & Pérusse and repeal their glossary of processes of numerical competence. *Brain and Behavioral Sciences, 11*, 600.
Thomas, R. K., Fowlkes, D., & Vickery, J. D. (1980). Conceptual numerousness judgments by squirrel monkeys. *American Journal of Psychology, 93*, 247–257.
Thompson, R. F., Mayres, K. S., Robertson, R. T., & Patterson, C. J. (1970). Number coding in association cortex of the cat. *Science, 168*, 271–273.
Thorpe, W. H. (1956). *Learning instinct in animals* (pp. 385–394). Cambridge, MA: Harvard University Press.
von Glasersfeld, E. (1982). Subitizing: The role of figural patterns in the development of numerical concepts. *Archives de Psychologie, 50*, 191–218.
Warren, H. C. (1897). Studies from the Princeton Psychological Laboratory, VI–VII (VI. The reaction time of counting). *Psychological Review, 4*, 569–591.
Wolters, G., van Kempen, H., & Wijhuizen, G. (1987). Quantification of small numbers of dots: Subitizing or pattern recognition? *American Journal of Psychology, 100*, 225–237.
Woodworth, R. S. (1938). *Experimental psychology*. New York: Holt.
Woodworth, R. S., & Schlosberg, H. (1954). *Experimental psychology*. New York: Holt.
Young, A. W., & McPherson, J. (1976). Ways of making number judgments and children's understanding of quantity relations. *British Journal of Educational Psychology, 46*, 328–332.

CHAPTER EIGHT

Quantitative Relationships Between Timing and Counting

Hilary A. Broadbent
Russell M. Church
Brown University

Warren H. Meck
B. Carey Rakitin
Columbia University

Duration and number are attributes of all discrete stimuli. The duration of a stimulus and the number of elements or segments of the stimulus typically covary, so it has been difficult to demonstrate that animals can discriminate between stimuli on the basis of number alone. Animals can discriminate between stimuli on the basis of duration, and both an information-processing model (Gibbon & Church, 1984) and a formal theory (Scalar Timing Theory) (Gibbon, 1981; Gibbon, Church, & Meck, 1984), successfully account for many of the facts of animal timing (Gibbon, 1991). There are now convincing data that animals can discriminate the number of stimuli, even when stimulus durations are uninformative. The chief purpose of this chapter is to review some data suggesting that the same mechanism is used for timing and counting, and to describe how timing models have been extended to account for number discrimination.

In a recent review of the literature on the ability of animals to count, Davis and Pérusse (1988) proposed definitions of various levels of numerical competence, drawing on the principles developed by Gelman and Gallistel (1978). These principles were originally intended to apply to counting in young children, but it is appropriate to apply them more generally. A second purpose of this chapter is to show how these principles apply to information-processing and connectionist models of animal timing and counting.

NUMBER DISCRIMINATION IS NOT TIME DISCRIMINATION

One can study an animal's ability to discriminate number in either of two ways: The animal can be required to produce a particular number of responses (response counting) or it can be required to respond to a stimulus on the basis of the number of elements or segments (stimulus counting). There is a problem with response-counting experiments because it is difficult to remove confounds of number with time. Mechner (1958) trained rats to respond a given number of times on the right lever of a two-lever box with the following procedure: After a fixed number of right-lever responses, a left-lever response was reinforced. However, there is an obvious difficulty with this procedure because time, rather than number, could be the controlling variable: The rat could respond at a relatively constant rate on the right lever for a particular amount of time, rather than for a particular number of responses. Mechner and Guevrekian (1962) and Laties (1972) attempted to deal with this problem by manipulating the rate of responding, but there is evidence that their manipulations also change the rate of the internal clock (Maricq, Roberts, & Church, 1981; Meck, 1983). Thus, it remains difficult to rule out temporal control of responding. Suppose, for example, that a treatment increases the rate of responding by 10% and also increases the rate of an internal clock by the same percentage. After the treatment, the physical time required to produce a given number of responses will be reduced by 10%, but the subjective time required to produce a given number of responses would not be affected by the treatment.

Another approach has been to impose a blackout of variable duration between responses (Wilkie, Webster, & Leader, 1979), so that the overall duration of the trial no longer provides the animal with a precise cue. However, the duration of the trial was related to the number of responses, and the possibility that the animals were timing the trial rather than counting their responses was not ruled out (Fernandes & Church, 1982).

It is easier to argue that responding is controlled by number in a stimulus-counting experiment than in a response-counting experiment. The temporal characteristics of stimuli can be precisely controlled to a degree that is impossible when animals count their own responses. Fernandes and Church (1982) attempted to demonstrate that number of stimuli could be the basis for a discrimination even when all time intervals were controlled. They trained rats to press the right lever of a two-lever box following a stimulus consisting of 2 white-noise segments, and to press the left lever following a stimulus consisting of 4 white-noise segments (as diagrammed in Rows a–c of Fig. 8.1). The number of segments, but not the stimulus duration, provided reliable information

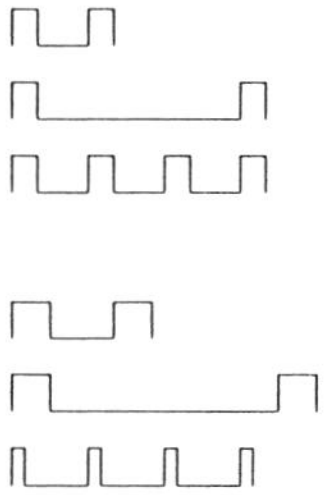

8. TIMING AND COUNTING

		Description of Stimulus		
		Number of segments	Stimulus duration(s)	Sum of segments (s)
a)	⎍⎍	2	1.2	.4
b)	⎍⎍	2	3.2	.4
c)	⎍⎍⎍⎍	4	3.2	.8
d)	⎍⎍	2	1.6	.8
e)	⎍⎍	2	3.6	.8
f)	⎍⎍⎍⎍	4	2.8	.4

FIG. 8.1. Segmented stimuli used to demonstrate that number discrimination does not require a correlated time cue.

regarding the response to be reinforced. During this training, the sum of segment durations also provided reliable information. However, in a subsequent unreinforced transfer test, the duration was changed, as diagrammed in Rows d–f of Fig. 8.1. The probability of a response was controlled by number of segments rather than the sum of segment durations. The animals discriminated between 2 and 4 segments, in spite of contradictory temporal cues. This indicates that number, not the overall duration of the stimulus nor the summed durations of the segments, controlled the behavior.

An alternative test of number as an adequate discriminative stimulus is to train animals with redundant stimuli (both time and number were reliably associated with the reinforced response), and to test one group with number constant and test the other group with time constant (Meck & Church, 1983). If both groups could perform the task, then training with time and number as redundant cues leads animals to learn both time and number. One training stimulus consisted of 2 half-second white-noise segments (as shown in Row a of Fig. 8.2), and the other consisted of 8 half-second white-noise segments (as shown in Row b of Fig. 8.2). After discrimination had been established, half the rats received stimuli that were intermediate in overall duration (4 s) and in total segment duration (2 s), but varied in the number of segments (2 to 8) (see Fig. 8.2, Rows c–i). The other half received stimuli that were intermediate in overall number of segments (4), but varied in overall duration (2 to 8 s) and total segment duration (1 to 4 s), as shown in Fig. 8.2, Rows j–p. In each case, the animals were able to discriminate among stimuli on the basis of the relevant dimension, whether it was number or duration. Thus, when number and duration are confounded dimensions, such that either pro-

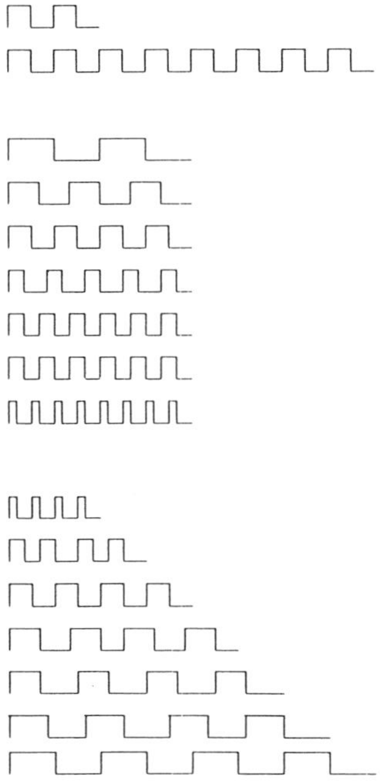

		Description of Stimulus		
		Number of segments	Stimulus duration(s)	Sum of segments (s)
a)	⊓⊔⊓⊔	2	2	1
b)	⊓⊔⊓⊔⊓⊔⊓⊔⊓⊔⊓⊔⊓⊔⊓⊔	8	8	4
c)	⊓___⊔___⊓⊔	2	4	2
d)	⊓⊔⊓⊔⊓⊔	3	4	2
e)	⊓⊔⊓⊔⊓⊔⊓⊔	4	4	2
f)	⊓⊔⊓⊔⊓⊔⊓⊔⊓⊔	5	4	2
g)	⊓⊔⊓⊔⊓⊔⊓⊔⊓⊔⊓⊔	6	4	2
h)	⊓⊔⊓⊔⊓⊔⊓⊔⊓⊔⊓⊔⊓⊔	7	4	2
i)	⊓⊔⊓⊔⊓⊔⊓⊔⊓⊔⊓⊔⊓⊔⊓⊔	8	4	2
j)	⊓⊔⊓⊔⊓⊔⊓⊔	4	2	1
k)	⊓⊔⊓⊔⊓⊔⊓⊔	4	3	1.5
l)	⊓⊔⊓⊔⊓⊔⊓⊔	4	4	2
m)	⊓⊔⊓⊔⊓⊔⊓⊔	4	5	2.5
n)	⊓⊔⊓⊔⊓⊔⊓⊔	4	6	3
o)	⊓⊔⊓⊔⊓⊔⊓⊔	4	7	3.5
p)	⊓⊔⊓⊔⊓⊔⊓⊔	4	8	4

FIG. 8.2. Segmented stimuli used to compare number and time discrimination.

vides an adequate basis for discrimination, animals learn both number and duration.

NUMBER DISCRIMINATION IS RELATED TO TIME DISCRIMINATION

Although animals can discriminate the number of events independently of any particular duration, they may use the same mechanism for the discrimination of time and number. Four quantitative similarities between time and number have been reported in a series of experiments in which animals were trained on the bisection procedure (Meck & Church, 1983).

First, rats were trained to discriminate a 2-s stimulus with 2 segments ("short/few," Row a of Fig. 8.2) from an 8-s stimulus with 8 segments ("long/many," Row b of Fig. 8.2). They were then tested with duration fixed at 4 s (Rows c–i) or number fixed at 4 (Rows j–p). The psychophysical function generated by plotting the probability of a "long" response as a function of the overall duration was indistinguishable from the function generated by plotting the probability of a "many" response as a function of the number of cycles. The superposition of the functions for time and number means that an index of the accuracy of the discrimination, such as the difference limen or a Weber fraction, would reveal equivalent discriminability on the basis of number or duration. In both cases, the point of subjective equality (the point at which 50% of the responses were "long" or "many") was at the geometric mean of the number or duration of the training signals. A similar result was obtained when rats were trained to make one response if the stimulus either consisted of 2 segments or if the stimulus duration was 2 s (Rows c and j), and to make another response if the stimulus either consisted of 4 segments or if the stimulus duration was 4 s (Rows i and p). Then, as in the previous experiment, the rats were tested with duration fixed at 4 s (Rows c–i) or number fixed at 4 (Rows j–p). Figure 8.3 shows the very similar psychophysical functions for time and number.

Second, methamphetamine produced a leftward horizontal shift in the psychophysical functions for both timing and counting. This treatment had previously been shown to produce a leftward shift in the psychophysical function for duration discrimination that is consistent with an increase in the speed of an internal pacemaker (Maricq & Church, 1983;

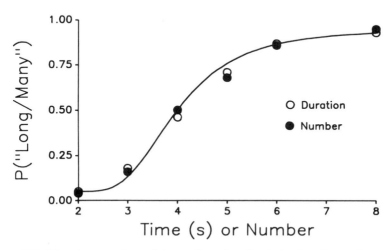

FIG. 8.3. A comparison of time and number discrimination. The probability of a "long" response as a function of the duration of continuous stimuli and the probability of a "many" response as a function of the number of segments of segmented stimuli.

Maricq, Roberts, & Church, 1981; Meck, 1983). The effect of the drug on number discrimination was equivalent to its effect on duration discrimination, suggesting a shared physiological basis for timing and counting.

Third, when a tactile stimulus (mild electric shock) was interposed between the noise segments, the animals responded with respect to the total number of events (auditory plus the tactile segments). The amount of cross-modal transfer was equivalent for counting and timing.

Fourth, if timing and counting are related processes, there should be some logical way of mapping one onto the other. This was accomplished by training rats to discriminate between continuous stimuli differing in duration, and then testing them on segmented stimuli differing in number. Rats were trained in a temporal bisection procedure using white-noise signals to press one lever following a 1-s continuous noise-stimulus and the other lever following a 2-s continuous noise stimulus, as shown in Rows a and g of the first column of Fig. 8.4. Continuous stimuli of intermediate durations (as shown in Rows b–f of the first column of Fig. 8.4) were introduced without reinforcement. When these animals were subsequently tested with segmented stimuli (as shown in the second column of Fig. 8.4), they treated 5 segments as being equivalent to 1 s of duration, and 10 segments as being equivalent to 2 s of duration (Meck, Church, & Gibbon, 1985). This equivalence was not affected by the duration of each segment (.5 or 1.0 s). The conclusion was that 5 segments were equiva-

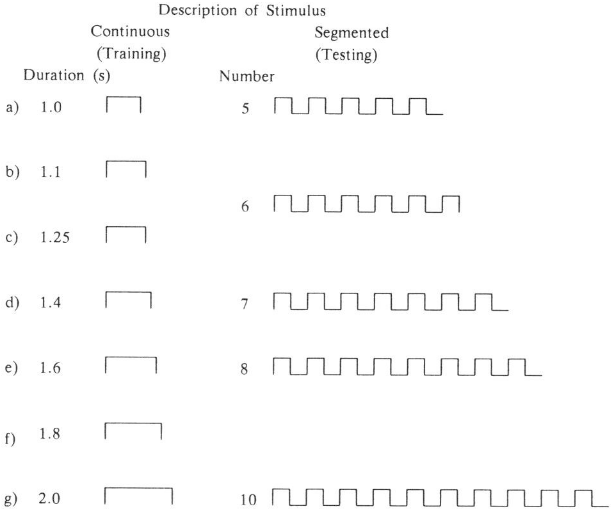

FIG. 8.4. Stimuli used to determine the relationship between the number of segments and the duration of a continuous stimulus.

8. TIMING AND COUNTING   **177**

lent to a duration of 1 s. In a previous experiment (Meck & Church, 1983), the same conclusion was made on the basis of a different range of continuous durations (2 to 4 s) and a different range of number of segments (10 to 20).

AN INFORMATION-PROCESSING MODEL OF TIMING AND COUNTING

If timing and counting have the same physiological bases, as suggested by the data, then any model that accounts for one should account equally well for the other. An information-processing model developed by Church and Gibbon (1982) and elaborated by Meck and Church (1983) has been very successful as an implementation of scalar timing theory (Gibbon, 1981) to account for animal timing (see top panel of Fig. 8.5). According to the model, there is an endogenous pacemaker that emits pulses at regular intervals. At the start of an interval to be timed, a switch is closed that gates pulses into an accumulator until the end of the interval, at which time the switch is opened. The number of pulses in the accumulator represents the duration of the interval. This quantity can be compared with a sample of remembered time of reinforcement retrieved from reference memory.

There is an assumption that the switch between the pacemaker and the accumulator can be operated in several different modes, and this assumption permits the extension of the timing model to counting. For example, suppose a stimulus consisted of several segments. The switch can be closed from stimulus onset to termination (run mode), or it can close only during each segment and open between the segments (stop mode). If the switch were operated in run mode, the number of pulses in the accumulator would equal the pacemaker rate times the duration of the signal. If the switch were operated in stop mode, the number of pulses in the accumulator would equal the pacemaker rate multiplied by the total duration of the segments. Animals can perform either run mode or stop mode timing, depending on the response contingencies (Roberts, 1981; Roberts & Church, 1978).

A third mode of timing is also possible. Rather than the switch being closed during the segments of an intermittent stimulus, as it is in stop mode, at each onset of a segment the switch can close for some fixed time and then open again. This is called event mode, and is shown in the bottom panel of Fig. 8.5. At the occurrence of each segment, the switch closes for a relatively fixed time. Thus, the increase in the number of pulses in the accumulator is a function of the number of segments, but independent of segment duration. The number of pulses in the ac-

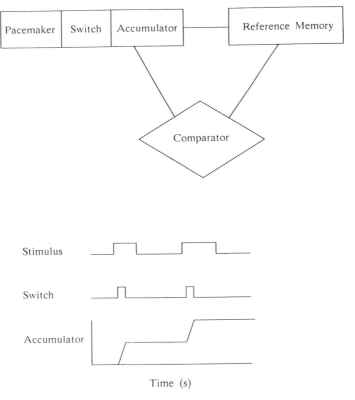

FIG. 8.5. An information-processing model of timing or counting (top panel). A switch operating in event mode: A fixed duration of closure at each segment onset (bottom panel).

cumulator at the end of the signal would represent not the total duration during which the signal was present, but rather some constant multiplied by the number of signal onsets. The event mode of operation of the switch would allow an animal to count successive events using the same mechanism that is used to time them (Meck & Church, 1983).

Each of the four findings that link timing and counting can be explained by the information-processing model with different switch modes for timing and counting. First, the model predicts that the psychophysical functions for timing and counting should be similar because the same time-scale (assumed to be linear) and comparison rule (assumed to be a ratio rule) are used for both tasks. Because the same pacemaker and the

8. TIMING AND COUNTING   179

same memory storage processes are used for both timing and counting, the variability in the representation of time and number are also predicted to be equivalent.

The second fact, that methamphetamine produces a leftward horizontal shift in the psychophysical functions similar in magnitude for timing and counting, is also predicted by the model. Suppose methamphetamine speeds up the rate of the pacemaker by 10%. Subjective time for a given duration with methamphetamine treatment would then be 10% longer than it would be without treatment, because more pulses would be gated into the accumulator during the same amount of physical time when the pacemaker was running faster. Thus, a leftward shift in the psychophysical function for time is predicted. The same principle applies when the clock is running in event mode: The fixed-duration burst at the start of each stimulus segment would contain more pulses under methamphetamine treatment than it would under normal conditions. Thus, a leftward shift in the psychophysical function for number is also predicted.

The third fact to be explained is the equivalent amount of cross-modal transfer of timing and counting. Because the same mechanism is used for both timing and counting, the only difference being the mode of switch operation, it is reasonable to assume that the rules that govern the use of the mechanism for timing the durations of stimuli of various modalities would also apply to counting stimuli of various modalities.

The fourth fact to be explained is the mapping of five counts onto 1 s of duration. In event mode, the switch closes for the same fixed amount of time at the onset of each stimulus segment. If that fixed amount of time equals 200 ms, then the number of pulses gated into the accumulator in five counts would be equal to the number of pulses gated into the accumulator in 1s of run-mode timing.

A CONNECTIONIST MODEL OF TIMING AND COUNTING

A connectionist model of timing has been described in detail elsewhere (Church & Broadbent, 1990). Briefly, the model consists of two banks of oscillators, one for storage of times and one for retrieval, instead of a single pacemaker (see Fig. 8.6). The oscillators are assumed to be tightly coupled so that all the oscillators vary from their mean rate by the same percentage on any given trial. For the sake of simplicity, it is assumed that the periods of the oscillators are related in 2:1 ratios, so that each oscillator runs twice as fast as the next slowest one.

Current subjective time is represented as a storage vector containing

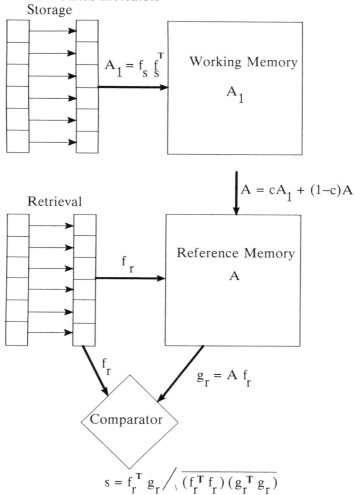

FIG. 8.6. A connectionist model of timing or counting. From Church and Broadbent (1990). Copyright © 1990 by Elsevier Science Publishers. Reprinted by permission.

8. TIMING AND COUNTING   **181**

information about the half-phases of the oscillators. It is filled with +1s and −1s that, because of the 2:1 coupling, constitute a binary representation of subjective time. (Note, however, that the individual elements of the storage vector are processed independently and in parallel, rather than as a binary number.) An autoassociative matrix storing the subjective time of reinforcement is formed by taking the outer product of the storage vector with its transpose. This matrix is referred to as *working memory*. *Reference memory* is a matrix formed by a linear combination of some percentage of the previous reference memory matrix plus a percentage of the current working memory matrix.

To determine whether or not a response should be made at a given time, the current subjective time, a retrieval vector similar in form to the storage vector can be compared with the times stored in the reference memory matrix as follows: The inner product of the reference memory matrix and the retrieval vector is an output vector. Roughly speaking, the greater the similarity between the retrieval vector and the output vector, the closer the current time is to the stored time of reinforcement. The measure of similarity that is used is the cosine of the angle between the retrieval vector and the output vector. If the cosine is greater than a specified threshold, a response is considered to have occurred.

There are two ways in which the connectionist model for timing could account for numerical discriminations. If the oscillators run for a fixed amount of time at the beginning of each segment of a discontinuous signal, then the values in the status indicators at the end of the series of segments would represent the number of segments. This is analogous to event-mode timing in the information-processing model, where pulses from the pacemaker are gated for a fixed amount of time into the accumulator and the value in the accumulator at the end of the series represents the total number of events. The number could then be stored in an autoassociative matrix, just as time is stored when the model is used for timing.

In the information-processing version, the pacemaker continues to run as events occur; it is the gating action of the switch that controls the bursts of pulses. There is a problem with the connectionist version of event-mode timing because it is perhaps not entirely plausible that an organisms's oscillators would run for a short period and then reset each time an event occurs. This difficulty can be avoided by allowing the system access to information about the phases of the various oscillators. For simplicity's sake, the current model makes use of half-phase information only, and the oscillators themselves are reset and restarted with each event.

There is another way in which the connectionist model for timing could be used to count a number of events. Number can be derived from temporal information by multiplying rate by time, or by dividing the overall duration of a series of stimulus segments by the mean intersegment interval. Although the connectionist model has no mechanism for carrying out multiplication or division of scalar (one-dimensional) quantities, a similar kind of information can be extracted from vector quantities that represent times.

Consider stimuli that differ in intersegment interval (c) and in overall duration (d), which have a constant number of segments (i.e., have a constant ratio of d to c). The upper left panel of Fig. 8.7 shows a stimulus with an intersegment interval of 1 s (c = 1) and an overall duration of 2 s (d = 2). These times can be represented in two separate vectors, with the elements either set to +1 (black) or -1 (white). (For convenience, Fig. 8.7 shows only four elements of each vector.) The separate vectors can be concatenated to produce a single longer vector, and the outer product of this vector with its transpose forms a working memory representation of the stimulus. The strength of the connection is set to +1 (black) when the corresponding elements are in the same state, and -1 (white) when the corresponding elements are in the opposite state. (The diagonal is set to zero on the assumption that each element is connected to each other element, but not to itself.)

If the intersegment interval (c) and the overall duration (d) were doubled, the number of segments would remain the same. The working memory representation of a 4-s stimulus with a 2-s intersegment interval is shown in the center panel of Fig. 8.7, and the working memory representation of an 8-s stimulus with a 4-s intersegment interval is shown in the right panel.

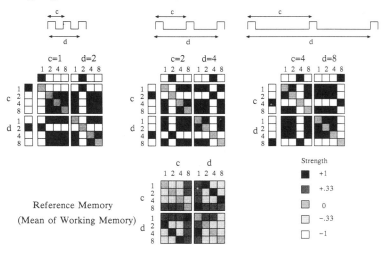

FIG. 8.7. A representation of a segmented stimulus in reference memory.

8. TIMING AND COUNTING   **183**

Reference memory was approximated simply as the mean of the three working memory components. The strength of connection is indicated by the shading. The dominant feature of this representation of reference memory is the diagonal stripe.

Deriving number from intersegment interval and overall duration has the desirable feature of linking time and number. In a more realistic example, it would be necessary to use more elements in each of the vectors and to add variability. Presumably, the addition of variability would add some noise, but it would permit appropriate generalization to occur (Church & Broadbent, 1991). A method for extracting the information in the stripes must be made explicit. This version of the connectionist model predicts that speeding up the clock with methamphetamine would produce leftward shifts in the psychophysical function for number discrimination, if the learning of the intersegment interval (which is shorter and more frequent than the overall stimulus duration) proceeds more rapidly than the learning of the overall stimulus duration. It also yields testable predictions. Although rats can count events that occur irregularly (Davis, Memmott, & Hurwitz, 1975), our analysis predicts that events that happen at regular intervals should be counted with greater accuracy.

CAN RATS COUNT?

The criteria of Gelman and Gallistel (1978) are generally accepted as a definition of "counting." Do the information-processing and connectionist models of counting meet these criteria?: (a) A unique tag must be applied to each object to be counted (the 1:1 principle); (b) these tags must be ordered consistently (the stable-order principle); and (c) the last tag in the set must stand for the set as a whole (the cardinal principle).

There are two additional principles that address the flexibility of counting rather than define it. The first of these two stipulates that the tags should be able to be used for any set of stimuli (the abstraction principle). The second states that stimuli should be countable regardless of the order in which they are presented (the order-irrelevance principle). An organism that could count one kind of object, but not others, would not have a very flexible counting system. However, it would be difficult to argue that the organism was not counting the one kind of object that was within its capabilities. In other words, just as one should examine numerical competence within the animal's range, so, too, one must consider that an animal may have a limited domain of counting, but that, within that domain, the animal may nonetheless have a reasonably complex representation of number.

The information-processing model's representation of number (Meck

& Church, 1983) fulfills the demands of the Gelman and Gallistel (1978) principles. The value in the accumulator after a given number of events uniquely represents that number, and therefore obeys the 1:1 principle. Furthermore, the value is arrived at by progressing through an ordered sequence of values, each of which uniquely represents a number. Thus, the stable-order principle is followed. The value in the accumulator at the end of the sequence stands for the sequence as a whole when it is compared with the value stored in reference memory, and therefore the model obeys the cardinal principle. Finally, there is nothing in the model that requires that the accumulator value be tied to any particular stimulus modality or set of items or events, or any particular order of presentation. Cross-modal transfer of numerical discrimination has been demonstrated (Meck & Church, 1983). Thus, the model adheres to both the abstraction and the order-irrelevance principles.

One version of the connectionist model for counting also adheres to the Gelman and Gallistel (1978) principles. The arguments regarding the event-mode version of the connectionist model are very similar to those for the information-processing model. The collection of values in the status indicators at the end of a number of events uniquely represents that number, and the model therefore obeys the 1:1 principle. The representation of each number is arrived at by passing through representations of lesser numbers along the way and are thus necessarily ordered, so that the stable-order principle is followed. Perhaps the easiest way to see the correspondence between the information-processing representation of number and the connectionist representation of number is to consider that the collection of values in the status indicators of the connectionist model can, assuming the oscillators are tightly coupled, be viewed as a binary number. This binary number equals the number represented by accumulated pulses in the information-processing version, assuming that the fastest oscillator of the connectionist version has the same period as the pacemaker of the information-processing version, and assuming that the length of time the oscillators run at each event in the connectionist version equals the length of time that the switch is open for each burst of pulses in the information-processing version. (Note again, however, that the collection of values in the status indicators is not processed as a binary number in the connectionist model; nor does the model necessarily require that the oscillators be tightly coupled.) As before, the abstraction and order irrelevance principles are concerned more with how the model is applied than with the actual mechanisms proposed, and as with the information-processing model, there is nothing in the connectionist model that precludes transfer to new stimulus modalities or new stimulus orders.

An interesting feature of the implementation of the connections model derives number from intersegment interval and overall duration: Although it does clearly provide a representation of number, it violates the 1:1 principle. In Fig. 8.7, the same number is encoded in each of the three working memory matrices. Thus, there is not a unique tag for each number. This implementation could not, therefore, be considered in the formulation of Gelman and Gallistel (1978) to be an instance of counting per se, although number is clearly represented in a way that would permit crossmodal transfer and insensitivity to order of presentation of stimuli.

CONCLUSIONS

A substantial body of evidence indicates that timing and counting have a shared mechanism (Church & Meck, 1984). Not only do some drug treatments have similar effects on the two processes (Maricq et al., 1981; Maricq & Church, 1983; Meck, 1983; Meck & Church, 1983) but timing and counting appear to map directly onto one another so that five counts is equivalent to 1 s (Meck & Church, 1983; Meck et al., 1985).

An information-processing implementation of scalar timing theory accounts well for the available data on counting and timing. However, the current implementation of this model has some awkward features. For example, in order to store multiple values in reference memory, multiple sample distributions must be established and maintained. Presumably, this would entail replications of the neural hardware. In addition, any single sample distribution must either increase indefinitely in size as more and more samples are stored, or some information must be irretrievably lost to make room for new samples.

A connectionist implementation of scalar timing theory, currently under development, also provides a possible explanation for the links between counting and timing. This model avoids some of the undesirable features of the information-processing model. In the current formulation of this model, reference memory is a set of associative matrices, each for a different type of information. Within each matrix, more than one value can be stored without replication of hardware (although, of course, the capacity for separable values is not infinite). Furthermore, an infinite number of samples of each value can be stored in a matrix without increasing the size of the matrix at all. Early information diminishes in importance, but is never lost altogether. However, the connectionist model has not yet been used to provide accurate fits to counting data.

Whichever model turns out in the long run to account better for the available data, it is clear that any model that provides an explanation of counting should also explain the data that link counting and timing. It is also clear that a model can follow the widely accepted definitional principles of Gelman and Gallistel (1978) without attributing unreasonable computational abilities to animals.

REFERENCES

Church, R. M., & Broadbent, H. A. (1990). Alternative representations of time, number, and rate. *Cognition, 37,* 55–81.

Church, R. M., & Broadbent, H. A. (1991). A connectionist model of timing. In M. L. Commons, S. Grossberg, & J. E. R. Staddon (Eds.), *Neural network models of conditioning and action* (pp. 225–240). Hillsdale, NJ: Lawrence Erlbaum Associates.

Church, R. M., & Gibbon, J. (1982). Temporal generalization. *Journal of Experimental Psychology: Animal Behavior Processes, 8,* 165–186.

Church, R. M., & Meck, W. H. (1984). The numerical attribute of stimuli. In H. L. Roitblat, T. G. Bever, & H. S. Terrace (Eds.), *Animal cognition* (pp. 445–464). Hillsdale, NJ: Lawrence Erlbaum Associates.

Davis, H., Memmott, J., & Hurwitz, M. B. (1975). Autocontingencies: A model for subtle behavioral control. *Journal of Experimental Psychology: General, 104,* 169–188.

Davis, H., & Pérusse, R. (1988). Numerical competence in animals: Definitional issues, current evidence, and a new research agenda. *Behavioral and Brain Sciences, 11,* 561–579.

Fernandes, D. M., & Church, R. M. (1982). Discrimination of the number of sequential events by rats. *Animal Learning & Behavior, 10,* 171–176.

Gelman, R., & Gallistel, C. R. (1978). *The child's understanding of number.* Cambridge, MA: Harvard University Press.

Gibbon, J. (1981). On the form and location of the psychometric bisection function for time. *Journal of Mathematical Psychology, 24,* 58–87.

Gibbon, J. (1991). Origins of scalar timing. *Learning and Motivation, 22,* 3–38.

Gibbon, J., & Church, R. M. (1984). Sources of variance in an information processing theory of timing. In H. L. Roitblat, T. G. Bever, & H. S. Terrace (Eds.), *Animal cognition* (pp. 465–488). Hillsdale, NJ: Lawrence Erlbaum Associates.

Gibbon, J., Church, R. M., & Meck, W. H. (1984). Scalar timing in memory. In J. Gibbon & L. G. Allan (Eds.), *Annals of the New York Academy of Sciences: Timing and time perception* (pp. 52–77). New York: New York Academy of Sciences.

Laties, V. (1972). The modification of drug effects on behavior by external discriminative stimuli. *Journal of Pharmacology and Experimental Therapeutics, 183,* 1–13.

Maricq, A. V., & Church, R. M. (1983). The differential effects of haloperidol and methamphetamine on time estimation in the rat. *Psychopharmacology, 79,* 10–15.

Maricq, A. V., Roberts, S., & Church, R. M. (1981). Methamphetamine and time estimation. *Journal of Experimental Psychology: Animal Behavior Processes, 7,* 18–30.

Mechner, F. (1958). Probability relations within response sequences under ratio reinforcement. *Journal of the Experimental Analysis of Behavior, 1,* 109–121.

Mechner, F., & Guevrekian, K. (1962). Effects of deprivation upon counting and timing in rats. *Journal of the Experimental Analysis of Behavior, 5,* 463–466.

Meck, W. H. (1983). Selective adjustment of the speed of internal clock and memory processes. *Journal of Experimental Psychology: Animal Behavior Processes, 9,* 171–201.

Meck, W. H., & Church, R. M. (1983). A mode control model of counting and timing processes. *Journal of Experimental Psychology: Animal Behavior Processes, 9,* 320–334.

Meck, W. H., Church, R. M., & Gibbon, J. (1985). Temporal integration in duration and number discrimination. *Journal of Experimental Psychology: Animal Behavior Processes, 11,* 591–597.

Roberts, S. (1981). Isolation of an internal clock. *Journal of Experimental Psychology: Animal Behavior Processes, 7,* 242–268.
Roberts, S., & Church, R. M. (1978). Control of an internal clock. *Journal of Experimental Psychology: Animal Behavior Processes, 4,* 318–337.
Wilkie, D. M., Webster, J. B., & Leader, L. G. (1979). Unconfounding time and number discrimination in a Mechner counting schedule. *Bulletin of the Psychonomic Society, 13,* 390–392.

PART THREE

COUNTING IN HUMANS AND ANIMALS: THEORETICAL PERSPECTIVES

CHAPTER NINE

Animal Number Abilities: Implications for a Hierarchical Approach to Instrumental Learning

E. John Capaldi
Purdue University

Consider an animal, which in order to gain a food reward, responds either by traversing a runway or completing a fixed ratio. It seems totally evident that the behavior involved in running from the start to the goal portion of a runway or completing a number of bar-presses or pecks is organized in the sense that a variety of subbehaviors occur in a particular order, an order suited to gaining reward. For example, the naive animal on being placed in the runway, sniffs the air, rears, walks back and forth, freezes, and otherwise roams about in a seemingly aimless manner. Contrast this ineffectual responding with that of the trained rat. It orients immediately to the start door. The start door opens, it races down the runway. As it approaches the goal cup it slows down. Finally it stops, positions itself over the goal cup, and eats if food is available. Let us refer to the various responses that, taken together, result in the animal being moved as it were from start to goal as the runway response. Sometimes, components of the runway response have been employed as the dependent variable (see, e.g., Miller, 1956).

Employing the runway response itself as the dependent variable is a commitment to the view, implicitly if not explicitly, that various of its subbehaviors, ranging from orienting to the start door to entering the goal box, are organized into a whole, what I call a *trial chunk*. Treating the runway responses as a coherent functional unit or chunk is justifiable because it obviously is one. That is, transformation from an ineffectual series of independent responses to an efficient mechanism for gaining reward is clearly evident, even to casual inspection. The ultimate or most important reason for selecting some behavior as our unit of analysis is because it is useful for some purpose, so I suggest that the runway response, important as it may be for some purposes, is not useful for others. Indeed, the runway response, which is clearly composed of subunits, is itself a constituent of a larger unit. Trials are organized into series. Larger units, *series,* are postulated because they are useful explanatory devices. Unfortunately, their operation is not open to casual inspection: It must be inferred. Recent data from my laboratory indicate that series are organized into still higher order units called *lists,* a matter discussed elsewhere (Capaldi, in press; Capaldi, Miller, Alptekin, & Barry, 1990).

Counting, as it is studied in my laboratory, involves using a series of reward events as a discriminative cue. For example, a rat might receive three food rewarded runway trials in a row followed by a nonrewarded trial. Trained in this manner, the rat runs rapidly on each of the rewarded trials and slowly on the nonrewarded trial (see, e.g., Capaldi & Miller, 1988a). Clearly this indicates, as becomes evident, that the rat is employing the prior three rewarded trials as a signal for nonreward. I am suggesting, then, that by investigating counting in my laboratory I am simultaneously investigating the organizational capacities of the rat in instrumental learning tasks. I have then a dual purpose in studying counting. I am interested in animal numerical abilities. I am also interested in demonstrating that understanding instrumental learning involves postulating a unit of analysis higher than either the components of a runway response or the runway response itself. That higher-level unit, of course, is a series of independent trials. I have long been guided by the assumption, which I hope is supported by observations contained in this chapter, that the ability of animals to employ number of events as a discriminative cue is important for understanding a variety of instrumental learning phenomena. In that sense, I have long been interested in animal number abilities. As an examination of some of my earlier work shows, I emphasized the ability of animals to encode number of events as an explanation for a variety of conventional instrumental learning phenomena (e.g., Capaldi, 1964, 1966, 1967). According to my view, then and now, the ability of animals to employ number of events as a discriminative cue has broad and deep implications for animal learning theory. In this chapter my intention is to relate animal number ability and animal learning theory in a way others may find useful.

Although I have long been committed to the idea that animals employ number of events as a discriminative cue, I began seriously to think about animal number ability, and more specifically animal counting, after having read the well-known and highly influential review of that literature

9. ANIMAL NUMBER ABILITIES   193

by Davis and Memmott (1982). All of us owe Davis and Memmott (1982) a debt of thanks for bringing up that topic to our collective attention. However, as I read their excellent review of the animal number literature I was struck by how much my views about the place of animal counting in the general scheme of things differ. In a nutshell, they believe that counting is not very important in an animal's day-to-day activities. They began by defining counting in a generally acceptable way, drawing on Gelman and Gallistel (1978). Counting involves applying distinctive tags (internal representations) in one-to-one correspondence to events. The tags should be applied in a stable order over occasions (stable-order principle), and any event may be assigned any tag, thus the order of presenting to be enumerated items is irrelevant (order-irrelevance principle). Sure, animals can count, they went on to say. But, they suggested, counting by animals is a rare thing, something animals engage in only when forced to do so, when all alternative means of solving a problem is denied them. I had, implicitly if not explicitly, come to a far different conclusion on the basis of the results of conventional learning investigations (see, e.g., Capaldi, 1964, 1966). It is a conclusion I intend to emphasize in this chapter. Animals, at least animals as highly developed as the rat, count routinely, I suggest. More important for present purposes, the ability of animals to use number of events as a discriminative cue, and perhaps even to count, has important implications for learning theory. That is to say, the view that animals can use number of events as a discriminative cue is compatible with certain popular theoretical approaches in learning and is incompatible with others. If this is indeed the case, then few other things will cause more people to devote energy, intelligence, and skill to the topic of animal numerical abilities, such as our concern with animal learning theory. My major concern in this chapter, then, is to place animal counting in the larger context of animal learning theory.

GENERAL OVERVIEW

I begin by describing a variety of conventional instrumental learning data that are compatible with the notion that animals are capable of employing number of events as a discriminative cue. I attempt to describe the knowledge structure that underlies performance in those situations. I attempt to show that other forms of popularly accepted knowledge structures are not consistent with these findings. This does not mean these alternative knowledge structures are useless; far from it. However, precisely what their function might be lies outside the scope of this chapter. Having examined conventional learning data, I turn to studies that have been run in my laboratory specifically concerned with animal numerical abilities. I attempt to show, first of all, that those studies indicate animals employ number of events as a discriminative cue. Put differently, I attempt to show that the data cannot be explained in terms of various alternatives to number. Subitizing is especially emphasized. Having shown that number is the controlling variable in those studies, I then attempt to show that the data indicate that the animals counted. These counting data, also, I suggest, are consistent with some theories of learning, but not others.

CONVENTIONAL LEARNING DATA

Types of Associations

Attempts to explain performance in instrumental learning tasks have led to the postulation of three types of association, each with subtypes. According to the classical view, antecedent stimuli (S) are associated with the instrumental response (R), an S-R association (e.g., Hull, 1943). With Hull's S-R view, the role of the reinforcer (S*) consequent on instrumental responding is simply to promote the formation of the S-R bond. The S-R view is at best incomplete. A variety of data, a very considerable variety, indicates that the reinforcer, S*, is part of what is learned (see, e.g., Capaldi, 1967; Colwill & Rescorla, 1986). Many hold the view that an association is formed between R and S*, an R-S* association (e.g., Bolles, 1972; Colwill & Rescorla, 1986; Konorski, 1948; Mackintosh & Dickinson, 1979; Tolman, 1933). The third type of association postulated is between S and S*, an S-S* association.

Consider now subvariates of these associations. According to one view, perhaps the most generally accepted single view in instrumental learning, S elicits a classically conditional form of the goal response (r_g), which produces a stimulus (S_c^*), which is associated with R, an S_c^*-R association (e.g., Amsel, 1958; Spence, 1956, 1960; Trapold & Overmier, 1972). The S_c^*-R association, which of course is merely a variety of S-R association, may be employed to explain a variety of phenomena that cannot be explained by the less elaborate S-R theory. It is important to keep in mind that S_c^* is ultimately dependent on classical conditioning established between S and the reinforcer.

Another view holds that the reinforcer is represented via memory (e.g., Capaldi, Nawrocki, Miller, & Verry, 1986). In this case, the memory of one or more prior reinforcers, S_m^*, can be employed to anticipate the current reinforcer, S_a^*, an S_m^*-S_a^* association. The greater the similarity between conditions at retrieval and at storage, the greater the tendency to retrieve S_m^*. The S_m^*-S_a^* association is a variety of the more conventional

9. ANIMAL NUMBER ABILITIES   **195**

S-S* association. We do not preclude the possibility that S_m^* may set the occasion for an R-S_a^* association or there may be S_m^*-R associations. Indeed, some data suggest the necessity of postulating some form of S-R association (see Mackintosh, 1983). However, our emphasis in this chapter is on S_m^*-S_a^* associations.

Reward Schedule Data

Consider rats that receive an alternating series of food rewarded (F) and nonrewarded (N) trials in a runway. Early in training under the alternating schedule, rats run faster following F trials (on N trials) than following N trials (on F trials). This highly inappropriate behavior is precisely what would be expected by two-factor theory. Following F trials, the S-r_g association must be stronger than following N trials. If trials were completely independent, by which I mean if events on one trial were not remembered on the next trial, there would be no reason to suppose that rats would ever abandon the tendency to run faster on N trials than F trials. Of course, rats do abandon this highly inappropriate form of behavior, even when trials are separated by as much as 24 hours (Capaldi & Lynch, 1966; Jobe, Mellgren, Feinberg, Littlejohn, & Rigby, 1977). Eventually the rats come to run faster on F trials than on N trials (pattern running). There would seem to be no way to derive this more appropriate form of responding, often called *pattern running*, on the basis of an S_c^*-R association. Nor can it be explained, it seems, on the basis of an R-S* association. If trials were independent, the animal would have no way of knowing that this particular R would be followed by F or N.

Pattern running is completely explicable in terms of S_m^*-S_a^* associations. Early in training, before the S_m^*-S_a^* association is formed, the rat responds in a manner appropriate only to S_m^*. Thus, if it remembers food reward it runs faster than if it remembers nonreward. With training the S_m^*-S_a^* association is formed and grows progressively stronger. Thus, the animal eventually learns that the memory of food reward signals nonreward (run slow), whereas the memory of nonreward signals food reward (run fast). It would be difficult to explain acquisition responding under the single-alternating schedule in the absence of postulating an S_m^*-S_a^* association. Although other associations may be involved, available data do not seem to allow us to come to a definite conclusion on this score. Alternation data suggest trials are united into a series. It is, of course, the S_m^*-S_a^* association that unites individual trips down a runway into a higher-level unit, a series.

Pattern running under the single-alternation schedule indicates that rats can remember whether or not they received food reward. Other data are consistent with the view that rats can remember how many rewarded trials and how many unrewarded trials occurred. It is such data that give rise to the suspicion that rats, in addition to employing number of events as a discriminative cue, may be able to count or enumerate such events as well. First, consider the number of nonrewarded trials. If percentage of reinforcement is held constant but the sequence of F and N trials is varied, the following will be found: As the number of N trials that precede F trials increases, resistance to extinction will increase (see, e.g., Capaldi, 1964, 1966; Haggbloom, 1980; Jobe et al., 1977). The explanation of this in terms of S_m^*-S_a^* associations is straightforward. In acquisition, the animal learns how many N trials in succession precede an F trial. This representation becomes a signal for F. In extinction, the animal compares the series of N trials currently in working memory with the associative representation stored in reference or long-term memory, which was formed in acquisition. As the two representations become increasingly dissimilar, that is, as the number of N trials in working memory exceeds the number of N trials signaling F in reference memory, responding progressively declines in vigor. Let us apply this view to certain representative findings. Under consistent reinforcement acquisition, no N trial directly signals an F trial, and consistent reward produces rapid extinction. If partial reinforcement is employed, but the memory of N is not allowed to become a signal for F by one means or another, extinction will be rapid (see, e.g., Capaldi & Spivey, 1963; Grosslight & Radlow, 1956; Haggbloom, 1982). And, of course, the longer the string of N trials preceding an F trial in acquisition, the greater the resistance to extinction (see, e.g., Capaldi, 1964, 1966; Capaldi, Miller, & Alptekin, 1989; Jobe et al., 1977). Recent data suggest that effects attributed to number of N trials, such as those already suggested, cannot be explained in terms of timing or associating position cues with F (Capaldi et al., 1989). No other theory has attempted to explain the findings described here. This is itself cause for comment. I suggest that where trials are united into series, behavior simply cannot be explained by S-S*, S-R, R-S*, or S_c^*-R associations.

The extinction analysis concerning the role of number of N trials preceding an F trial supplied is fully supported by acquisition data. Consider an experiment reported by Capaldi and Verry (1981, Experiment 5). Rats received two series of trials on an irregular basis in the runway. Trials of a series were separated by about 30 s, series by about 20 min. One series began with a 20-pellet reward, which was followed by four trials, each of which terminated in nonreward, a 20-0-0-0-0 series. The other series was 0-0-0-0-20. The series occurred irregularly, so the animal could not correctly anticipate whether Trial 1 would terminate in 20 pellets or 0 pellets. But once the first trial of the series was received, the reward outcomes on all remaining trials could be correctly anticipated,

9. ANIMAL NUMBER ABILITIES

provided the animal could remember the reward outcomes on all prior trials. That is, at any point in either series, except for Trial 1, the animal could anticipate the current reward outcome by remembering all prior reward outcomes. The results of the experiment, which can be seen in Fig. 9.1, indicate that on each trial of either series the animals were able to determine what events had occurred and what events were to occur. By the end of training, the animals ran appropriately in each series. On Trial 1, which was irregularly reinforced, the rats ran rapidly. On Trial 2, nonreinforced in both series, the animals ran slowly. In the 20-0-0-0-0 series, the animals ran slowly on all trials subsequent to Trial 1. In the 0-0-0-0-20 series, however, the rats speed of running progressively increased from Trial 1 to Trial 5, until by Trial 5 speed was very rapid. Capaldi & Verry (1981) said of this finding, "Rats remembered how many nonrewarded events there had been accurately enough to suggest that they were using some form of counting mechanism" (p. 44). The acquisition data shown in Fig. 9.1 are completely consistent with the view that rats have an accurate notion of precisely how many nonrewarded trials precede a food reinforced trial, an inference first drawn on the basis of extinction data (e.g., Capaldi, 1964, 1966). In Fig. 9.1 it seems clear that over the trials of the 0-0-0-0-20 series, the animal's expectancy of reward increased. It is hard to imagine within the context of two-factor theory how r_g could increase in strength over a series of nonrewarded trials. Thus S_c^*-R associations seem inadequate to explain the behavior of the animals in the 0-0-0-0-20 series. Nor can the behavior of the animals in the 0-0-0-0-20 series be explained in terms of R-S* associations. If trials are independent, the animal would have no way of determining if this particular response was to terminate in reward or nonreward. Thus, the assumption that trials are not independent is required to explain a wide variety of reward schedule data, both in acquisition and extinction (see also Capaldi & Miller, 1988c). Although I do not review the findings here, memory of reward events has also been shown to be a factor regulating performance in discrimination tasks (e.g., Capaldi, Berg, & Morris, 1975; Haggbloom, 1980, 1988).

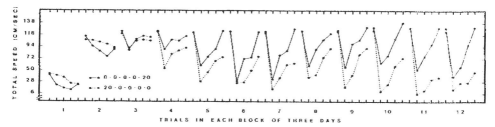

FIG. 9.1. Total speed in blocks of 3 days on each of the five trials of each of the series 0-0-0-0-20 and 20-0-0-0-0 in Phase 1.

We have seen that S_c^*-R associations and R-S^* associations cannot explain a wide variety of acquisition and extinction findings that are favorable to the view that animals form S_m^*-S_a^* associations. The opposite, however, is not the case. There are certain classes of data, I include among these devaluation findings (e.g., Colwill & Rescorla, 1986; Dickinson, 1989), differential outcome findings (e.g., Peterson, 1984), and so-called classical conditioning-instrumental conditioning interactions (e.g., Trapold & Overmier, 1972), which strongly support the view that the reinforcer is part of what is learned in various instrumental learning situations. The process by which the reinforcer is represented, is often said to be classical conditioning. Let me be clear. In all of the previously cited studies, the reinforcer could have been represented via memory. And it could have been represented via classical conditioning. Both alternatives are viable because in none of the studies was any attempt made to test classical conditioning versus memory as the process responsible for the reinforcer representation.

Let me elucidate the major issue, as I see it, with reference to the S_m^*-S_a^* association. The appearance of S_m^* occurs via memory retrieval. The S_m^*-S_a^* association may be achieved via classical conditioning, although there is reason to doubt this (e.g., Goodall & Mackintosh, 1987; Holman & Mackintosh, 1981). That is, the animal learns the relationship between a particular S_m^* and a particular S_a^*. For example, the animal learns under a single alternating series of rewarded and nonrewarded trials that the memory of reinforcement is followed by nonreinforcement and vice versa. The theoretical choice then is between S_m^*, a reinforcer representation produced via memory, and S_c^*, a reinforcer representation produced via classical conditioning. If it happens that differential outcome studies, devaluation studies, and so-called classical-instrumental interaction studies produce results more consistent with S_c^* than S_m^*, that would present us with quite a paradox. As I have shown, a variety of reward schedule data support S_m^* and not S_c^*.

Basic Procedure

In the prior section we examined studies showing that animals could employ number of nonrewarded events as a signal for either food reward, F, or nonreward, N, either in acquisition or extinction. In studies from my laboratory explicitly concerned with counting, number of F trials have been employed as signals for F or N. The basic procedure, employed in all studies, may be illustrated employing the two series, FFFN and NFFFN. Each animal received both series on an irregular basis, trials of a series

9. ANIMAL NUMBER ABILITIES   **199**

separated by about a 30-s intertrial interval, series presentations being separated by about a 10-min interval. By FFFN, for example, is meant that Trials 1, 2, 3, and 4 terminate in F, F, F, and N, respectively. And, of course, by NFFFN is meant that Trials 1, 2, 3, 4, and 5 terminate in N, F, F, F, and N, respectively. Trained under these two series (see Capaldi & Miller, 1988b, Experiment 4), rats learned to anticipate their terminal N trial of both series by running slower on that trial than on any prior trial. Consider why this finding suggests that animals employed number of F events as the discriminative cue on the terminal N event. Although number of F events was a consistent signal for the terminal N event in the two series, a variety of other events were not, and so these events could not have supported discriminative responding. In the two series, the terminal N event was preceded by different numbers of trials (3 vs. 4), different numbers of trips down the runway (3 vs. 4), different numbers of intertrial intervals (3 vs. 4), different numbers of goal box confinements (3 vs. 4), and so on. The time elapsing from the onset of Trial 1 to the onset of the terminal N event also differed in the two series; as did the sum of the confinement time in the goal box, the sum of the response times, and so on. These differences were achieved by the simple expedient of preceding one of the series by N, the NFFFN series. In both the NFFFN series and the FFFN series, the terminal N trial is preceded by three F events. Our interpretation of discriminative responding under the FFFN and NFFFN series is as follows. Zero prior F events was a signal for F or N on Trial 1, because series were presented irregularly. One prior F event was a signal for F, as were two prior F events. But three prior F events was a signal for N.

Before the study described can be taken to implicate number of F events as the discriminative cue, several other matters must be considered. Before dealing with those matters, however, let us pause briefly to consider the implications of the findings for various approaches to learning based implicitly, if not explicitly, on the view that trials are independent. In classical conditioning, trials involving CS-US pairings are assumed to strengthen the association between the two. Thus, it seems entirely anomalous from the standpoint of two-factor theory that responding would decease in strength following three F trials. Moreover, from the viewpoint of R-S* associations, the animal would have no way of determining which of its responses was to terminate in F and which in N. Nor can S-S* associations explain the findings. Thus, the results of counting studies considered in this section are as unfavorable to S_c^*-R, R-S*, and S-S* associations as are the results of reward schedule studies considered in the prior section.

SUBITIZING AND OTHER ALTERNATIVES TO NUMBER

We have seen that the procedure of employing two series such as FFFN and NFFFN rules out a variety of alternatives to number of F trials as a discriminative cue. But others remain. For example, the rats might be using the amount of food ingested as a discriminative cue. Or they might be timing the intertrial intervals elapsing between F trials only. To evaluate these and other alternatives, a variety of steps were taken by Capaldi and Miller (1988a) in various studies they reported. For example, following the development of fast running to all trials but the terminal N trials of the series, FFN and NFFN, intertrial intervals were increased or decreased in temporal duration, as were confinement times in the goal box of F trials. These manipulations rule out various temporal cues as the basis of discriminative responding, for example, summing the confinement duration in the goal box on F trials. In another experiment, rats were trained under the series F'FFN intermixed with a single N trial. Presentation of the F'FFN series and the N trial were separated by about 10 min. F' and F are meant to represent two qualitatively different food rewards. Having learned to anticipate the terminal N trial of the F'FFN series, the animals were shifted either to the series FFN and NFFN or FFFN and NFFFN. In both cases, the animals required only one or two presentations of the shift series to correctly anticipate all trials, that is, run fast on all trials but the terminal N trials of both series. I offer a counting interpretation of these findings and related findings later. For now, I just want to say that the shift findings clearly cannot be interpreted in terms of discriminative cues arising from amount of food ingested, time elapsing between F trials, and so on.

Subitizing

Subitizing refers to the rapid assignment of a numerical tag to a small number of simultaneously presented items, usually one to four. According to one view, small numbers of simultaneously presented items are not specifically enumerated or counted but, rather, are perceived or apprehended much as we perceive, say, color (see, e.g., Gelman & Gallistel, 1978; Mandler & Shebo, 1982; Woodworth, 1938). Subitizing is generally understood to be a nonnumerical process. Thus, to say an animal has subitized is to say an animal has not employed number at all, let alone counted. There is disagreement as to whether developmentally counting grows out of subitizing or vice versa (see Gelman & Gallistel, 1978). Of major interest here are the frequently expressed notions that subitizing is a nonnumerical process and the results of many experiments interpreted in terms of number or counting can be interpreted in terms of subitizing. Despite the fact that subitizing is itself not well understood, and indeed may be the result of a prior counting process, it is frequently invoked at the drop of a hat as an alternative explanation of experiments represented as supporting counting. Perhaps because of the Clever Hans debacle, the notion seems to be current that if any conceivable alternative explanation of a counting experiment can be imagined, then the results may be summarily dismissed as evidence for a numerical process. Subitizing fits this bill admirably.

Numerous animal investigations employing simultaneous presentation of items have been said to involve subitizing by Davis and Pérusse (1988). Furthermore, they have extended the rather vague notion of subitizing from its original base of simultaneously presented items, where some evidence of a sort exists for it in humans (see Miller, chap. 7, this vol.), to successively presented items, where unfortunately no evidence of any sort can be found. As Capaldi and Miller (1988b) suggested, in extending the notion of subitizing from simultaneous to successive presentation of items, Davis and Pérusse (1988) seem to be unconstrained in their speculations. Be that as it may, they believe that items presented successively may have a rhythm or cadence that can be and is learned. No experimental evidence was presented in support of this proposition. Davis and Pérusse (1988) then went on to assume that items presented successively can be more easily learned on a rhythmic basis than on the basis of enumeration. This chain of reasoning was then applied to results from my laboratory. According to this view, animals trained under, for example, the FFFN and NFFFN series, learned to anticipate the terminal N event on a rhythmic basis rather than by enumerating F events. For a variety of reasons mentioned by Capaldi and Miller (1988b), I regard it as most unlikely that our findings can be interpreted reasonably in terms of learning a rhythm, even assuming that it is easier to learn rhythm than to employ number, which is itself by no means clear. In any event, Davis and Pérusse suggested (personal communication, July 1988) that if animals trained under the two series, for example, FFN and NFFN ran faster on all trials except the terminal N trials when the intertrial interval in original training was varied, it would support our counting interpretation and not their rhythm view. Accordingly, I decided to run an experiment to settle the matter. The purpose of the unpublished experiment was to determine if under the conditions employed in my laboratory, F events are more likely enumerated or subitized.

Recently in our laboratory, we employed two groups of rats in a runway, both of which received the series FFN and NFFN in original learning. Group C received a constant 30-s intertrial interval between all trials of both series. Group V received a varied intertrial interval, consisting of an equal number of 20 s, 30 s, and 40 s, intertrial intervals averaging out to 30 s. To consider extremes, Group V received an NFFN series in which all trials were separated either by a 20-s interval or by a 40-s intertrial interval. At other times, the animals of Group V might experience all three intertrial intervals, 20 s, 30 s, and 40 s, in the series NFFN, the interval between Trials 1 and 2 being, say 20 s, that between Trials 2 and 3 being, say, 40 s and that between Trials 3 and 4 being, say, 30 s. Similar procedures, of course, were employed in connection with the FFN series received by Group V. Except for the differences mentioned, all procedures employed in the experiment under consideration were highly similar, if not identical, to those employed in our prior studies (see, e.g., Capaldi & Miller, 1988a). Following training of the sort described earlier for 40 days, both groups were now trained under the variable intertrial interval procedures for 4 days. Consider the implications of the experiment.

If Group V learns at all, it would suggest that our procedure can be employed to study animal counting. However, if Group V learns more slowly than Group C, it would suggest that it is easier to learn a rhythm than to count. If both groups learn equally well, it would suggest, however, that animals count in preference to employing the sort of rhythm cues speculated on by Davis and Pérusse (1988). Finally, if Group C, having learned to respond properly in original learning, breaks down completely in shift, it would suggest that where a rhythm as defined by Davis and Pérusse (1988) is present, counting is completely overshadowed. But if Group C is completely unaffected by the shift, it would suggest that rhythm is an inconsequential factor in our experiments, and perhaps generally.

The results of the experiment suggest, unequivocally, that rhythm learning is an inconsequential factor in our experiments. First, aside from the fact that Group C tended to run a little more slowly overall than Group V, the two groups differed little in the original learning phase. Slower running by Group C was due entirely to a single slow animal, which nevertheless ran more slowly on the terminal N trials than on the F trials. In addition, when Group C was shifted, it continued to respond under the varied intertrial intervals as it had under the constant intertrial interval. Figure 9.2 shows the mean running speed of each of the two groups on the F, F and N trials collapsed over series on the last 3-day blocks of trials in original learning and on each 1-day block of trials in shift. The unconnected points in Fig. 9.2 represent running speed on the N trial of the NFFN series. As may be seen, both groups ran rapidly to the initial N trial of the NFFN series and to the F trials of the series, while running slowly to the terminal N trial of the series. At no point, either in original learning or in shift, did the two groups differ significantly.

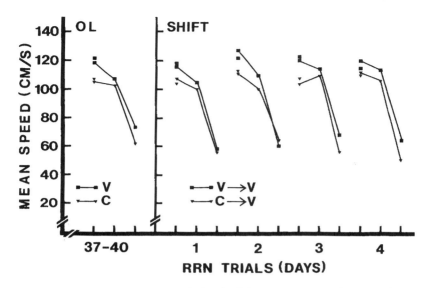

FIG. 9.2. Mean running speed of each of the two groups on the F, F and N trials collapsed over series on the last 3-day block of trials in original learning and on each 1 day block of trials in shift.

EVIDENCE FOR COUNTING

Let me briefly review the nature of the evidence examined so far. Conventional reward schedule investigations have provided a variety of acquisition and extinction data consistent with the idea that animals employ number of goal events as a discriminative cue. Such evidence is of an inferential sort. It has the following characteristic. If one assumes that number of, say, nonrewards was employed as a discriminative cue, then the results of many experiments, for which there is no other explanation, can be understood. This rather indirect sort of evidence has been supplemented recently by studies that explicitly employed number of goal events as a discriminative cue. These studies have provided more direct evidence indicating that number of goal events can be employed as a discriminative cue. That is, alternatives to number as a discriminative cue appear to have been ruled-out in such studies. We now examine evidence suggesting that animals can employ number of events as a discriminative cue as a result of an explicit counting or enumeration process.

By a number tag I mean an internal marker that corresponds to number of events. Humans employ conventional number tags, for example, one, two, three, and so forth. The nature of animal number tags is unknown. However, we can determine how they function. Before we are willing to conclude that some animal is capable of counting, we might demand that its use of number tags pass certain tests. The number of tests we might decide to employ may turn out to be very large and various. Some animals may be able to pass more of these tests than others. No doubt many such tests remain to be devised. Some of the tests devised in our laboratory are considered here. We begin by examining a test that, if failed by any animal, would suggest its use of number is too primitive to be called counting.

It seems clear that many numerical discriminations are conditional in nature; the number of events to be enumerated being dependent on some other cue. For example, one might be asked to enumerate items with respect to location, that is, how many such and such here, and how many there. Another conditionality might be to enumerate items by type, for example, how many apples and how many oranges. Rats have passed both of these tests. Burns and Sanders (1987) provided rats in a runway of one brightness with the series FFN and NFFN and in a runway of another brightness with the series FFFN and NFFFN. In both runways, rats ran rapidly on all trials except the terminal N trials. Capaldi, Miller, and Alptekin (1989) obtained similar results when the count was conditional on type of food, either F or F'. Thus, rats provided with four different series, FFN, NFFN, F'F'F'N and NF'F'F'N ran rapidly on all trials except the terminal N trials. It seems clear that rats can conditionalize their count either on some cue such as alley brightness or the type of item being enumerated.

In Experiment 7 reported by Capaldi and Miller (1988a), four qualitatively different food items were employed, call them F, F', f, and f'. In original training, rats received the series FFN and F'FFN on odd days and the series F'F'N and FF'F'N on even days. Note, to respond appropriately in original training—that is, i.e., to run fast on all trials but the terminal N trials—the rats had to employ the following rule: Run slowly only after having received two food rewards of the same type. Rats responded appropriately to original training. The rats were subsequently shifted to the series ffN and f'ffN on odd days and the series f'f'N and ff'f'N on even days. To respond appropriately in shift, rats had to apply the rule learned in original training to completely new food items. When shifted, rats displayed appropriate behavior to the new series immediately. Thus the rats not only employed a complicated rule in original training, they generalized it to new food items in shift. These results indicate that the number tags employed by rats are general.

Consider two letters X and Y. If our task was to count letters we would have "two." If our task was to count Xs and Ys, we would have one

9. ANIMAL NUMBER ABILITIES   **205**

of each. And, of course, if our task was to count Xs, Ys, and letters simultaneously, our count would be "one X, one Y, and two letters." In sum, members of our species are capable of classifying events simultaneously as same and different for purposes of enumerating them. The series of experiments to be described indicate that rats possess the capacity to classify events as same or different for purposes of enumeration. The extent to which this capacity is developed in rats remains to be determined; we may have merely scratched the surface so far. I have already referred to Experiment 5 of Capaldi and Miller (1988a), in which rats trained F'FFN and N in original learning showed appropriate behavior almost immediately when shifted to either the series FFN and NFFN (count to two) or the series FFFN and NFFFN (count to three). The shift results are understandable on the following basis. In original learning under the series F'FFN, the rats counted both to two (F' differs from F and F) or to three (F', F and F are all similar). Capaldi and Miller (1988b) provided rats with a seemingly more difficult version of the same-different counting problem. We attempted to bias the animals to count only to two by initially training them F'FFN and FFN. The idea was that the FFN series might result in the enumeration of only the F events in the F'FFN series. Not so. On being shifted to FFFN and NFFFN the animals displayed appropriate behavior almost immediately. Moreover, when subsequently being shifted again, this time to the series FFN and NFFN, the animals again displayed appropriate behavior almost immediately.

Burns and Gordon (1988), employing procedures similar to ours, trained three groups of rats. As in our experiments, a group trained F'FFN and N showed appropriate behavior when shifted either to FFN and NFFN or to FFFN and NFFFN. The interesting additional finding reported by Burns and Gordon (1988) was poor transfer to FFFN and NFFFN by a group trained FFN and N and poor transfer to FFN and NFFN by a group trained FFFN and N. These results demonstrate that appropriate responding in shift following F'FFN and N training cannot be ascribed to various nonspecific factors associated with prior training. Rather, transfer was due to rats enumerating an F'FFN series as two Fs and three "food rewards."

On the basis of results reported in this section it is reasonable to assume that rats are capable of employing number of events as a discriminative cue by specifically enumerating them. We have seen that rats can conditionalize their count on the basis either of some external cue such as runway brightness, or the event being enumerated. We have seen that rats having learned to run slowly only after two reinforcers of the same kind were received (FF or F'F' but not FF' or F'F) transfer this rule immediately to two new food events and so ran slow after ff or f'f' but not after ff' or f'f. Such transfer indicates that rats can apply their number tags to novel events and they can do so according to a rule, that is, run slow only after two reward events of the same kind. Finally, we saw that rats trained either F'FFN and N or F'FFN and FFN transferred rapidly to a count-to-two problem (FFN and NFFN) or a count-to-three problem (FFFN and NFFFN), and such transfer was not due to various nonspecific factors associated with prior training. These findings are clearly consistent with the criteria for counting proposed by Gelman and Gallistel (1978). That is, rats applied number tags to items, one tag to each item (one-to-one principle) the tags were applied to items in the same order over occasions (stable-order principle), and any event could be assigned any tag (order-irrelevance principle). Finally, the tags employed by rats could be assigned to entirely novel items according to a specific rule, indicating that the tags are general or abstract and can be applied according to an abstract rule.

Counting Studies: What is Learned?

Rats trained F'FFN and N could employ any number of cues to predict the terminal N event of the F'FFN series. For example, they could have predicted the terminal N event by employing the time elapsing from the beginning of the F' trial to the beginning of the terminal N trial. Or the rats could have summed the time required to complete three runway responses. None of these cues, nor many others, which were confounded with number of rewarded events in the F'FFN series, would have enabled the animals to transfer immediately to either the FFN and NFFN series or the FFFN and NFFFN series. Transfer in this instance, as in other instances from our laboratory, is consistent with two ideas. First, rats are capable of counting—in this case they counted rewarded events. Moreover, rats are capable of categorizing two different reward events, F' and F, either as similar or different for purposes of enumeration. A third conclusion available from these and other findings from our laboratory bears emphasis. Despite the fact that under the F'FFN and N series rats could predict the terminal N event on many bases, they nevertheless counted reward events. This clearly indicates that counting is not some esoteric activity engaged in by rats when no other means of solution is open to them, as suggested by Davis and Memmott (1982) and Davis and Pérusse (1988). Rather, rats count routinely. By this I mean that it is reasonable to assume that rats count in a wide variety of conventional learning situations. I suggest that to fail to assume that rats are employing number of events as a discriminative cue in some learning situation may be to cut off an appropriate avenue of explanation.

When rats employ number of successively presented reward events as a discriminative cue they are employing a level of organization larger than the trial or single runway response. Trials are organized into series. Where a series of events is a discriminative cue, particularly a counting series such as F'FFN, interpretations stressing S-S*, S_c^*-R, or R-S* associations are simply not viable. There is a major problem with such views because they employ as the functional unit of analysis the behavior occurring on a trial. Examples include completing a fixed-ratio or variable-ratio schedule or traversing a runway from start to finish. Another approach is required. A representation sensitive to the number of prior events is indispensable. Memory is required to fill that role. We need an S_m^*-S_a^* association. Still another assumption is required. Counting is itself a cognitive activity in which a tag, a number tag, is applied to a variety of different events. We have seen that rats having learned to supply a particular number tag to one type of reward event—for example, F can apply that tag immediately to a different reward event, for example, F'. Number tags are clearly concepts, the type called classical concepts (Smith & Medin, 1981). Interestingly, rats can employ number concepts according to rules that are themselves concepts. Thus, rats are able to categorize, for example, F' and F either as similar or different for purposes of counting. And trained under the FFN, F'FFN, F'F'N, and FF'F'N series, they can learn that the discriminative cue is two events of the same kind and they can generalize that rule immediately to the new series, ffN, f'ffN, f'f'N, ff'f'N. Thus, it appears that rats are capable of representing reward events at a general level.

REFERENCES

Amsel, A. (1958). The role of frustrative nonreward in noncontinuous reward situations. *Psychological Bulletin, 55,* 102–119.

Bolles, R. C. (1972). Reinforcement, expectancy, and learning. *Psychological Review, 79,* 394–409.

Burns, R. A., & Gordon, W. V. (1988). Some further observations on serial enumeration and categorical flexibility. *Animal Learning & Behavior, 16,* 425–428.

Burns, R. A., & Sanders, R. E. (1987). Concurrent counting of two and three events in a serial anticipation paradigm. *Bulletin of the Psychonomic Society, 25,* 479–481.

Capaldi, E. J. (1964). Effect of N-length, number of different N-lengths, and number of reinforcements on resistance to extinction. *Journal of Experimental Psychology, 68,* 230–239.

Capaldi, E. J. (1966). Partial reinforcement: A hypothesis of sequential effects. *Psychological Review, 73,* 459–477.

Capaldi, E. J. (1967). A sequential hypothesis of instrumental learning. In K. W. Spence & J. T. Spence (Eds.), *The psychology of learning and motivation* (Vol. 1, pp. 67–156). New York: Academic Press.

Capaldi, E. J. (in press). Levels of organized behavior in rats. In W. K. Honig & G. Fetterman (Eds.), *Cognitive aspects of stimulus control.* Hillsdale, NJ: Lawrence Erlbaum Associates.

Capaldi, E. J., Berg, R. F., & Morris, M. D. (1975). Stimulus control of responding in the early trials of differential conditioning. *Learning and Motivation, 6,* 217–229.

Capaldi, E. J., & Lynch, D. (1966). Patterning at a 24-hour ITI: Resolution of a discrepancy more apparent than real. *Psychonomic Science, 6,* 229–230.

Capaldi, E. J., & Miller, D. J. (1988a). Counting in rats: Its functional significance and the independent cognitive processes that constitute it. *Journal of Experimental Psychology: Animal Behavior Processes, 14,* 3–17.

Capaldi, E. J., & Miller, D. J. (1988b). Number tags applied by rats to reinforcers are general and exert powerful control over responding. *The Quarterly Journal of Experimental Psychology, 40B,* 279–297.

Capaldi, E. J., Miller, D. J., & Alptekin, S. (1989). A conditional numerical discrimination based on qualitatively different reinforcers. *Learning and Motivation, 20,* 48–59.

Capaldi, E. J., Miller, D. J., Alptekin, S., & Barry, K. (1990). Organized responding in instrumental learning: Chunks and superchunks. *Learning & Motivation, 21,* 415–433.

Capaldi, E. J., Nawrocki, T. M., Miller, D. J., & Verry, D. R. (1986). Grouping, chunking, memory, and learning. *The Quarterly Journal of Experimental Psychology, 38B,* 58–80.

Capaldi, E. J., & Spivey, J. E. (1963). Effect of goal box similarity on the aftereffect of nonreinforcement and resistance to extinction. *Journal of Experimental Psychology, 66,* 461–465.

Capaldi, E. J., & Verry, D. R. (1981). Serial order anticipation learning in rats: Memory for multiple hedonic events and their order. *Animal Learning and Behavior, 9,* 441–453.

Colwill, R. M., & Rescorla, R. A. (1986). Associative structures in instrumental learning. In G. H. Bower (Ed.), *The psychology of learning and motivation* (Vol. 20, pp. 55–104). New York: Academic Press.

Davis, H., & Memmott, J. (1982). Counting behavior in animals: A critical evaluation. *Psychological Bulletin, 92,* 547–571.

Davis, H., & Pérusse, R. (1988). Numerical competence in animals: Definitional issues, current evidence, and a new research agenda. *Behavioral and Brain Sciences, 11,* 561–615.

Dickinson, A. (1989). Expectancy theory in animal condition. In S. B. Klein & R. R. Mowrer (Eds.), *Contemporary learning theories: Pavlovian conditioning and the status of traditional learning theory* (pp. 279–308). Hillsdale, NJ: Lawrence Erlbaum Associates.

Gelman, R., & Gallistel, C. R. (1978). *The child's understanding of number.* Cambridge, MA: Harvard University Press.

Goodall, G., & Mackintosh, N. J. (1987). Analysis of the Pavlovian properties of signals for punishment. *Quarterly Journal of Experimental Psychology, 39B,* 1–22.

Grosslight, J. H., & Radlow, R. (1956). Patterning effect of the nonreinforcement-reinforcement sequence in a discrimination situation. *Journal of Comparative and Physiological Psychology, 49,* 542–546.

Haggbloom, S. J. (1980). Reward sequence and reinforcement level as determinants of S-behavior in differential conditioning. *Animal Learning and Behavior, 8,* 424–428.

Haggbloom, S. J. (1982). Effects of N-R transitions during partial reinforcement pretraining of resistance to discrimination. *Animal Learning and Behavior, 10,* 61–64.

Haggbloom, S. J. (1988). The signal-generated partial reinforcement extinction effect. *Journal of Experimental Psychology: Animal Behavior Processes, 14,* 83–95.

Holman, J. G., & Mackintosh, N. J. (1981). The control of appetitive instrumental responding does not depend on classical conditioning to the discriminative stimulus. *Quarterly Journal of Experimental Psychology, 48,* 305–340.

Hull, C. L. (1943). *Principles of behavior.* New York: Appleton-Century-Crofts.

Jobe, J. B., Mellgren, R. L., Feinberg, R. A., Littlejohn, R. L., & Rigby, R. L. (1977). Patterning, partial reinforcement, and N-length as a function of reinstatement of retrieval cues. *Learning and Motivation, 8,* 77–97.

Konorski, J. (1948). *Conditioned reflexes and neuron organization.* New York: Cambridge University Press.

Mackintosh, N. J. (1983). *Conditioning and associative learning.* Oxford: Clarendon Press.
Mackintosh, N. J., & Dickinson, A. (1979). Instrumental (Type II) conditioning. In A. Dickinson & R. A. Boakes (Eds.), *Mechanisms of learning and motivation: A memorial volume to Jerzy Konorski* (pp. 143–169). Hillsdale, NJ: Lawrence Erlbaum Associates.
Mandler, G., & Shebo, B. J. (1982). Subitizing: An analysis of its component processes. *Journal of Experimental Psychology: General, III,* 1–22.
Miller, G. A. (1956). The magical number seven plus or minus two: Some limits on our capacity for processing information. *Psychological Review, 63,* 81–97.
Peterson, G. B. (1984). How expectancies guide behavior. In H. L. Roitblat, T. G. Bever, & H. S. Terrace (Eds.), *Animal cognition* (pp. 135–148). Hillsdale, NJ: Lawrence Erlbaum Associates.
Smith, E. E., & Medin, D. C. (1981). *Categories and concepts.* Cambridge, MA: Harvard University Press.
Spence, K. W. (1956). *Behavior theory and conditioning.* New Haven, CT: Yale University Press.
Spence, K. W. (1960). *Behavior theory and learning.* Englewood Cliffs, NJ: Prentice-Hall.
Tolman, E. C. (1933). Sign-Gestlat or conditioned reflex? *Psychological Review, 40,* 246–255.
Trapold, M. A., & Overmier, J. B. (1972). The second learning process in instrumental learning. In A. H. Black & W. F. Prokasy (Eds.), *Classical conditioning II: Current theory and research* (pp. 427–452). New York: Appleton-Century-Crofts.
Woodworth, R. S. (1938). *Experimental psychology.* New York: Holt.

CHAPTER TEN

A Conceptual Framework for the Study of Numerical Estimation and Arithmetic Reasoning in Animals

C. R. Gallistel
University of California, Los Angeles

The study of numerical discriminations and arithmetic reasoning in animals is a profoundly interesting topic in animal cognition. It illustrates abstract categorization and conceptualization in animals in a way that may have lessons to teach us regarding the nature of categorization and conceptualization in general. It may also be a primitive and foundational aspect of animal mentation, one on which other processes build. The purpose of this chapter is to sketch the conceptual framework within which I believe studies of numerical abilities in animals can most profitably be analyzed. The framework derives largely from the modern mathematical understanding of the number concept and of the means for establishing numerical reference.

Numbers are prototypic instances of categories. The number 3 represents an infinitely diverse set of collections with only one thing in common, their numerosity. Numerosity itself has no sensory attributes and the sensory attributes of the things that compose sets of numerosity "3" are in no way restricted by the numerical categorization process, so numerical categories are often taken to be maximally abstract. Empiricist theories of mind treat categories as abstractions from elementary sense data. The more abstract a category, the less elementary sense content it has. Empiricist theories of mind regard more abstract categories as more derived, farther removed from the foundations of mental activity, and hence characteristic of advanced stages of mental evolution and mental development. By contrast, rationalist theories of mind regard some of these same categories as the elementary or primitive foundations of mental activity. I argue that the experimental demonstrations of animal facility with numerical categorization and arithmetic reasoning, which I briefly review, is evidence in favor of the rationalist position.

THE CATEGORY-CONCEPT DISTINCTION

At the heart of the conceptual framework is the distinction between numbers as categories and numbers as concepts. A mental category, in the usage I propose, is a mental/neural state or variable that stands for things or sets of things that may be discriminated on one or more dimensions but are treated as equivalent for some purpose. The mental category corresponding to 3 is a mental state or variable that can be activated or called up by any set of numerosity 3 and is activated or called up for purposes in which the behaviorally relevant property of a set is its numerosity. On the other hand, a concept, in the usage I propose, is a mental/neural state or variable that plays a unique role in an interrelated set of mental/neural operations, a role not played by any other symbol. The numerical concept 3 is defined by the role it plays in the mental operations isomorphic to the operations of arithmetic, not by what it refers to or what activates it.

When we think about numbers, the concept of 3, like the concept of any other number, plays a unique role. Three is the only number that is the sum of 2 and 1; the only number that is the difference of 7 and 4, the smallest prime number greater than 1, the only prime divisor of 9, and so on. One, on the other hand, is the only number that when multiplied by any number yields that number. It is this unique property of the number 1 that makes it the multiplicative identity element in the number system. Similarly, 0 is the only number that when added to any other number yields that number, which makes it the additive identity element in the system. Each number in the infinite system of numbers has infinitely many roles (or propositions) like this that are true of it and it alone. Specifying one of these unique roles defines a number. This definition, the conceptual definition of a number, makes no reference to anything outside the system of arithmetic propositions. This definition defines a number in the same way that the game of chess defines its pieces. It does not in any way depend on references to anything outside the game.

One of the insights of the formalist approach to mathematical concepts is that mathematical concepts mean what they mean not by virtue of the categories of nonmathematical objects to which they refer, but rather by virtue of the roles they play in mathematical propositions, just as a chess piece is what it is not by virtue of how it looks (you can use a pebble

10. A CONCEPTUAL FRAMEWORK

for a knight) nor by virtue of what it refers to (no amount of church history will enlighten one about the role of a bishop), but rather by virtue of the role that it plays, the moves it makes. What this insight about the nature of numbers qua concepts means can perhaps be made clear by the following. Suppose we create a large set of plastic chips, each chip distinguishable from every other, and we secretly assign each chip to a unique number: the pink square chip is the number 0; the green round chip, the number 1; the purple triangular chip, the number 2; and so on. Suppose further that we create three slot machines that take these chips as inputs, two chips at a time. One machine is the adding machine. When a pair of chips is inserted into its slots, it spits out a third chip corresponding to the number that is the sum of the numbers corresponding to the two chips put in. The second machine is the multiplying machine. It spits out a chip corresponding to the number that is the product of the numbers corresponding to the chips put in. The third machine is the ordering machine. It gives a pleasing gong if the number corresponding to the chip put into the left slot is greater than or equal to the number corresponding to the chip put into the right slot; otherwise, it gives a raspberry sound. We can hand the chips and the three machines to mathematically sophisticated players, without telling them which chips correspond to which numbers nor which machine is which. They will be able to tell us which chips correspond to which numbers and which machine corresponds to which arithmetic operation simply by observing what happens when various pairs of chips are put into the machines. (I already indicated how to spot the additive and multiplicative identity chips, the chips corresponding to 0 and 1.) Thus, the chips (the numbers) are defined by their behavior in the symbol processing system, not by anything to which they refer, and likewise for the machines (the arithmetic operations).

The distinction between numbers as categories and numbers as concepts provides us with two separate experimental questions regarding the numerical competences of animals: (a) Do they have numerical categories? (b) Do they have numerical concepts?

They have numerical categories if they can be shown to respond to the numerosity of sets independent of any other attributes of the set. They have numerical concepts if they can be shown to have in their heads the equivalent of adding and subtracting machines, multiplying and dividing machines, and ordering machines.

More specifically, one set of questions (numerical categorization questions) are of the following form: Can an animal be taught to respond to a set of three items independent of the attributes of the items that compose the set or the manner in which they are arranged in space and time? If so, then the animal can be said to have or be able to form the category 3.

There is also a second set of questions (numerical concept questions):

Does the animal treat sets belonging to the category 3 as greater than sets belonging to the category 2 and less than sets belonging to the category 7? Does it anticipate that combining one set belonging to the category 2 with another set belonging to that category yields a set belonging to the category 4? Does it judge that the number by which one must multiply 2 to get 4 is the same as the number by which one must multiply 4 to get 8? If so, then it can be said to have the concepts 2, 3, 4, 7, and so on, and also to have the mental/neural equivalents of ordering, addition, and multiplication machines.

EXPERIMENTAL EVIDENCE

Evidence of Numerical Categorization

There is an extensive experimental literature demonstrating that the common laboratory animals discriminate serially and simultaneously presented sets on the basis of their numerosity, regardless of the sensory attributes of the items/stimuli composing the sets and regardless of the temporal and spatial patterns in which they are presented. There are also several experimental demonstrations that animals trained to make a discrimination on the basis of the numerical category of a set react appropriately to novel instantiations of the required numerosity. Other chapters in this volume cover these experiments at length. They have also been reviewed from the current perspective by Gallistel (1989, 1990). Here, I mention only a few of the most pertinent studies.

Counting Responses. Mechner (1958) and Platt and Johnson (1971) developed paradigms in which there is a criterion number (N) of presses on one lever. When that criterion is met, the rat must (at least on some percentage of the trials) break off pressing the initial lever and either press another lever (Mechner paradigm) or enter a feeding area monitored by a photocell (Platt and Johnson paradigm) in order to activate a food dispenser. The probability of the rat's breaking off a run of presses on the initial lever to try the food dispenser is plotted as a function of the number (n) of presses in the run. The modal value of n, the number of presses at which the probability of breaking off is maximal, increases as a linear function of the N, the required number of presses. Indeed, when the payoff matrix in these experiments is appropriately adjusted, the modal n equals N, for values of N at least as great as 24.

This is evidence for a discrimination based on the numerosity of the set of responses in a run, provided it can be shown that the discrimination is not based on a variable that covaries with n. The most obvious such variable is the temporal duration of the run. Mechner and Guevrekian (1962) ruled this out by manipulating the level of hunger. Varying hunger varied the response rate and hence the duration of a run of a given n by a factor of two without affecting the discrimination. In these discrimination data, the standard deviations in the probability distributions are large and they increase as the modal n increases. This means there is a fair probability that the rat will confuse 4 with the numerosity to either side of it. As the number gets larger, this tendency to confuse a numerosity with adjacent or nearby numerosities increases. We may conclude that the process by which the animal categorizes sets on the basis of their numerosity is a noisy one. It is as if the rat represented numerosity as a position on a mental number line (that is, a continuous mental/neural quantity or magnitude), using a noisy mapping process from numerosity to values on this continuum, so that one and the same numerosity would be represented by somewhat different mental magnitudes (positions on the number line) on separate occasions.

In the Mechner and Platt and Johnson paradigms, the animal makes a go–no go decision based on a count of the number of responses it has made. The experimenter does not control this number. In the paradigm developed by Rilling and McDiarmid (1965), the experimenter determines which of two numbers of responses the animal has just made, and the animal must then use that number as a discriminative stimulus in choosing which of two responses to make. The Rilling and McDiarmid apparatus confronts a pigeon with a row of three keys, which may be illuminated from behind. First the center key is illuminated. The pigeon must peck it until its light extinguishes and the two flanking keys light up. This happens after one of two experimenter-specified numbers of pecks at the center key. One of these numbers was fixed at 50. The other was varied in the course of the experiment. For the sake of illustration, assume that the value of this alternative number is 45. When the two flanking keys light up, the pigeon has made either 50 presses on the center key or 45. If it has made 50, then it is rewarded for pecking the key on the right flank; if 45, for pecking the left flank. When Rilling and McDiarmid varied the alternate number from 35 to 47, they found that the percentage choice of the correct key did not fall below 60% until the alternate number reached 47. Rilling (1967) showed that this discrimination was not based solely on the duration of the run of pecks to the center key. He contrasted the results when the duration of the run was in fact the independent variable with the results when the number was the independent variable and did a clever analysis showing that the pigeon was more likely to rely on number in lieu of duration than vice versa.

Counting Successively and Simultaneously Presented Stimulus Sets. Fernandes and Church (1982) showed that rats can discriminate the number of light flashes or tone beeps in a sequence of flashes or beeps and there is immediate cross-modal transfer of this discrimination, as might be expected if the numerosity rather than the sensory quality of the stimulus sequence is the decisive attribute from the animal's standpoint. They presented rats with a sequence of either 2 or 4 sounds just prior to inserting two levers into the test chamber. If the sequence had 2 elements, the rats had to choose the right lever to get a food reward; if 4, the left. The durations of the sounds and the intervals between them varied so that neither the duration of a single sound, nor the duration of an interval between sounds, nor the duration of the sequence could reliably lead to correct discrimination. For all the sequences, the rats chose correctly 70% to 90% of the time. When light flashes were substituted for sound beeps, the discrimination transferred. In a follow-up experiment, Davis (1984) showed that rats could discriminate a sequence of 3 beeps from both sequences of 2 and sequences of 4 beeps.

Meck and Church (1983) taught rats to choose between two levers on the basis of the sound sequence they had just heard. During training, the number of on–off cycles in the sequence and the duration of the sequence covaried. On subsequently interpolated unrewarded test trials, either sequence duration was held constant while the number of cycles varied or the number of cycles was held constant while the duration of the sequence varied. The results showed that during the training, when numerosity and duration covaried, the rats had learned both the duration and the numerosity and could use either with equal facility as a basis of discrimination when the other was held constant. Meck and Church also interpolated unrewarded test trials in which the stimulus was 4 noise bursts and 4 mild (tickling) shocks to the feet. The rats chose the "8" lever in response to this novel instantiation of the numerosity 8.

Church and Meck (1984) taught rats to respond to the left lever in response either to 2 flashes or 2 noise bursts and to the right lever in response to 4 of either stimulus. On unrewarded test trials interpolated among the training trials, they presented for the first time 2 flashes and 2 noise bursts (4 stimuli in all) or 1 flash and 1 noise burst (2 stimuli in all). The rats chose the right (four) lever in response to 2 flashes combined with 2 noise bursts, even though the left (two) lever was the rewarded response for each constituent of this novel compound stimulus. This is a very striking demonstration of the rat's tendency to focus on the numerosity of a set of stimuli, independent of the other qualities of the items enumerated (the sensory qualities of the stimuli and the spatiotemporal pattern of their presentation). In a conflict between aggregate numerosity and the numerosity of the subsets of homogeneous items, the rat goes with aggregate numerosity, even though the aggregate numerosity is instantiated by a novel combination of stimuli in a novel spatiotemporal pattern.

10. A CONCEPTUAL FRAMEWORK 217

The extensive studies of Capaldi and his collaborators on rats' counting the numbers of rewarded and unrewarded trials in a runway are reviewed elsewhere in this volume. Two aspects of these results should be stressed. The rats count events separated by substantial and variable temporal intervals (measured in minutes). The rats routinely transfer the discrimination from whatever rewards are used in training to novel rewards (Capaldi & Miller, 1988).

Boysen and Berntson (1989), Matsuzawa (1985), and Rumbaugh, Savage-Rumbaugh, and Hegel (1987) all showed that chimpanzees can learn to discriminate the numerosities of small sets presented as spatial arrays. Matsuzawa showed that the discrimination transferred to novel instantiations of the numerosity (5 blue gloves in the test set, following training with sets composed of 5 red pencils). Pepperberg (1987) showed the same thing in an African grey parrot and Davis (1984) showed it in the raccoon.

As this short review indicates, and as is extensively documented elsewhere in this volume, there is a large body of experimental literature showing that a variety of mammals and birds categorize stimulus sets on the basis of their numerosity and generalize immediately to novel instantiations of a given numerosity.

Is Counting the Categorization Process?

The processes or mechanisms by which animals and humans categorize natural and artificial kinds is one of the deep mysteries of psychology. When a pigeon is trained to peck only at slides showing trees, it discriminates novel tree slides as readily as it does the training exemplars (Herrnstein, 1990). What the process is that enables the mind of the pigeon to recognize what different instances of trees have in common is, to say the least, obscure. No one has yet been able to devise a process or mechanism that can categorize trees simply on the basis of static two-dimensional images. Attempts to construct such a process are an important endeavor in computer science. One might suppose that the process for classifying sets on the basis of their numerosity was even more mysterious and much of the discussion of this issue in the literature makes it seem mysterious. In fact, however, counting processes are all that are required to categorize sets on the basis of their numerosity. Counting processes are trivial to specify formally and to implement mechanically or electronically or (presumably) neurally.

In the Gelman and Gallistel (1978) analysis, counting processes or mechanisms map from numerosities to numerons. Numerons are representatives of numerosity, not necessarily verbal or written. Counting processes obey three constraints: (a) The *one-one principle*: Each item in the set being counted is assigned one and only one numeron (one and only one potential representative of the numerosity of a set); (b) the *stable ordering principle*: The order in which numerons are assigned is always the same. Note that it is the order in which numerons are assigned, not the order in which items are counted, that must be the same from one count to the next; (c) the *cardinal principle*: The final numeron assigned, and only the final numeron, is used as the representative of the numerosity of the set. Any process or mechanism that maps from numerosities to representatives of numerosity in accord with these principles is a counting process or a counting mechanism.

Meck and Church (1983) formulated and experimentally tested a model of the counting mechanism in animals. Their model of the counting mechanism is a simple modification of their model of the mechanism by which the animal generates representatives of temporal intervals. In their interval timing mechanism, a switch closes at the onset of an interval, gating pulses from a constantly running pulse generator (like the clock in a computer) to an accumulator. So long as the switch is closed, the contents of the accumulator grow linearly. When the switch opens, the magnitude in the accumulator is read into memory, where it serves as the representative of the duration of the interval the animal experienced. Meck and Church made this mechanism into a counting mechanism simply by assuming that when operating in the counting mode (or "event" mode, as they term it), the switch closes for a fixed duration for each item in the set being counted. When the switch has gated a burst of pulses to the accumulator once for each item in the set, the magnitude in the accumulator (the sum of the bursts) is read into memory, where it serves as the representative of the numerosity of the set, just as the magnitude (height) of a bar in a histogram represents a number of occurrences.

The mechanism Meck and Church proposed maps from to-be-counted events to states of the integrator, which are read into memory to serve as representatives of numerosity. Thus, states of the integrator constitute the numerons in this counting system. The mapping is one-one, because each event gates one and only one burst of pulses to the integrator; hence, each successively counted entity is paired with a successively higher quantity in the integrator. The states of the integrator are run through in an order that is always the same because the ordering relation for quantity or magnitude (that is, for the successive states of the integrator) is the same as the ordering relation for numerosity. The fact that adding successive increments to a quantity produces an ordered set of quantities just as adding successive ones produces an ordered set of numbers is part of the reason that quantity or magnitude can be represented numerically, and vice versa. Finally, the state of the integrator at the end of the series of events is taken to represent a property of the series (its

10. A CONCEPTUAL FRAMEWORK   **219**

numerosity), not a property of the final event itself. Thus, the mechanism conforms to the cardinal principle.

Church and Meck (1984) reviewed the ingenious experiments they did to support their model of the counting mechanism and, in particular, their contention that the counting mechanism is intimately related to the mechanism for timing the duration of intervals. Thus, there is a clearly spelled out and experimentally supported model of the counting mechanism by which animals categorize sets on the basis of their numerosity. On the other hand, it has often been suggested that animals determine at least small numerosities by a subitizing process whose defining property is a negative one, namely, that it is not a counting process (Davis & Pérusse, 1988). The persistent belief that small numerosities may be perceived by humans and animals without the mediation of a counting process does not appear to be empirically motivated.

Evidence for Arithmetic Reasoning

For us to grant that an animal makes use of numerical categories, we need evidence that it discriminates between sets on the basis of their numerosity. There is fairly abundant experimental evidence on this score. For us to grant that an animal has numerical concepts, we need evidence that it combines numerons in mental operations that are isomorphic to the small set of combinatorial operations that define the system of arithmetic ($\geq$, $+$, $-$, $*$, $\div$). The directly relevant experimental evidence on this score is much less extensive, but there are some strong results.

Perhaps the most compelling result comes from the experiment by Boysen and Berntson (1989) in which they taught a juvenile female chimpanzee to label sets of appropriate numerosity with the Arabic numerals 1 through 5. When she had learned this, she was able without further training to check the oranges concealed in two of three different hiding places and then choose the Arabic numeral that represented the total number of oranges. This by itself is not clear evidence of mental addition, because the result could have been mediated entirely by counting. The chimp may have counted the first set, then continued its count when it came to the second set, thus obtaining a representative of the total numerosity by counting across the two sets. An operation isomorphic to addition must combine one representative of a numerosity (the first addend) with a second representative of a numerosity (the second addend) to yield a third representative of numerosity (the sum). When one counts across sets, the second addend does not exist, because at no point in the process is there a representative of the numerosity of the second set. This problem—the possibility of "pseudo-addition" by counting across sets—confounds the interpretation of several experiments that might be taken to demonstrate mental addition and subtraction (e.g., Rumbaugh et al., 1987). However, Boysen and Berntson went on to replace the sets of oranges with numerals. Their chimp, without further training, chose the numeral that was the sum of the numerals she had seen at the two locations. This is exceptionally direct evidence of mental addition in a nonhuman animal.

Less direct but potentially compelling evidence for arithmetic operations with representatives of numerosity comes from experimental investigations of the decision process in number discrimination tasks. In the Mechner (1958) paradigm and Platt and Johnson's (1971) adaptation of it (reviewed earlier), the rat's probability of abandoning the lever to seek food peaked at a fixed difference between the target value (the required number of presses) and the current count (the number of presses since initialization of the counter). The size of this difference—the number by which the n at peak exceeded the required N—was a systematic function of the prematurity penalty. This implies that the rat's decision to try the feeder was determined by a computation of the difference between a reference numeron in memory and the numeron from its current count of its presses.

On the other hand, in the Meck and Church (1983) paradigm, where the rat has been trained to choose one lever after seeing or hearing one number of cycles and the other level after seeing or hearing two to four times as many cycles, when the rat is presented with an intermediate number of cycles, it chooses on the basis of the ratio between the number of test cycles and the two reference numbers, which is why the equipreference point is at the geometric mean (equiratio point) between the two reference values. This implies the computation of the ratio of two numerons.

Gibbon and his collaborators (Church & Gibbon, 1982; Gibbon, 1981; Gibbon & Church, 1981, 1990; Gibbon, Church, & Meck, 1984) studied the decision process in timing behavior in detail, amassing strong experimental evidence that this process involves adding, differencing, and dividing representatives of temporal intervals. Thus, there is extensive experimental evidence that the brain performs the basic arithmetic operations with the representatives of magnitudes (in this case temporal intervals). As already mentioned, these same authors (Church & Meck, 1984; Meck & Church, 1983; Meck, Church, & Gibbon, 1985) provided fascinating experimental support for their hypothesis that numerosity is represented by the same kinds of magnitudes (accumulator contents) that represent temporal durations. There is every reason to suppose that the brain is capable of applying to these mental magnitudes the same basic arithmetic operations whether the magnitudes stand for temporal intervals or for numerosities.

10. A CONCEPTUAL FRAMEWORK 221

The evidence that the brains of pigeons and rats routinely add, subtract, and divide with representatives of temporal duration and the close link between the processes that represent duration and the processes that represent numerosity remind us that one should not confine the search for mental operations indicative of a number concept to purely numerical tasks. Representatives of numerosity are also used conceptually when they are combined arithmetically with representatives of magnitudes, as for example, when the rate of event occurrence is calculated by dividing a numeron representing the number of observed events by a quantity that represents the duration of the interval of observation (a "duron," that is, a representative of duration). There is an extensive literature showing that a variety of animals when confronted with alternative foraging patches in which food occurs at different rates match the relative amounts of time they spend in the patches closely to the relative rates of food occurrence (see Gallistel, 1990, chap. 11 for review). Attempts to model this matching to rate by nonrepresentational associative processes have not been successful (Lea & Dow, 1984). It is reasonable to regard the abundant literature on rate or probability matching as evidence that animals compute rates in the obvious way—by dividing a number by an interval. Insofar as one is persuaded that such a computation underlies rate matching, one accepts that animals have a number concept.

CONCLUSIONS

In discussing the numerical abilities of animals, it helps to keep in mind the distinction between numbers as categories and numbers as concepts. Numbers are categories insofar as they stand for diverse sets that have in common only their numerosity. In other words, numbers are categories defined by the process that maps from sets to representatives of the numerosity of those sets. The only well-understood processes that make this mapping for sets of more or less arbitrary composition and for various spatiotemporal patterns of set presentation are counting processes. There is extensive experimental evidence that animals respond to sets on the basis of their numerosity, for both simultaneously and sequentially presented sets, and for sets of diverse composition, including sets composed of novel (untrained) items or presented in a novel (untrained) spatiotemporal pattern. Evidence of numerical discrimination extends up to set sizes of 50. The upper limit on the ability to discriminate sets on the basis of their numerosity has yet to be established. The literature on numerical discrimination thus gives us reason to think that animals categorize sets on the basis of their numerosity. Their ability to do so is probably mediated by a brain process that conforms to the three principles that define counting processes—the one-one principle, the stable ordering principle, and the cardinal principle.

Numbers are concepts insofar as they enter into operations isomorphic to the simple combinatorial operations that define the system of arithmetic. In general, the best place to look for evidence of these combinatorial operations is in the study of the decision processes in number discrimination tasks. The modest number of relevant studies suggests that the common laboratory animals order, add, subtract, multiply, and divide representatives of numerosity. Thus, they use numbers as both categories and concepts. Their ability to do so is not surprising if number is taken as a mental primitive—just as it is a scientific primitive—rather than as something abstracted by the brain from sense data only with difficulty and long experience.

ACKNOWLEDGMENTS

Parts of this chapter closely follow chapter 10 of Gallistel (1990).

REFERENCES

Boysen, S. T., & Berntson, G. G. (1989). Numerical competence in a chimpanzee (*Pan troglodytes*). *Journal of Comparative Psychology, 103,* 23–31.

Capaldi, E. J., & Miller, D. J. (1988). Counting in rats: Its functional significance and the independent cognitive processes which comprise it. *Journal of Experimental Psychology: Animal Behavior Processes, 14,* 3–17.

Church, R. M., & Gibbon, J. (1982). Temporal generalization. *Journal of Experimental Psychology: Animal Behavior Processes, 8,* 165–186.

Church, R. M., & Meck, W. H. (1984). The numerical attribute of stimuli. In H. L. Roitblatt, T. G. Bever, & H. S. Terrace (Ed.), *Animal cognition* (pp. 445–464). Hillsdale, NJ: Lawrence Erlbaum Associates.

Davis, H. (1984). Discrimination of the number three by a raccoon (*Procyon lotor*). *Animal Learning and Behavior, 12,* 409–413.

Davis, H., & Pérusse, R. (1988). Numerical competence in animals: Definitional issues, current evidence, and a new research agenda. *Behavioral and Brain Sciences, 11,* 561–615.

Fernandes, D. M., & Church, R. M. (1982). Discrimination of the number of sequential events by rats. *Animal Learning and Behavior, 10,* 171–176.

Gallistel, C. R. (1989). Animal cognition: The representation of space, time and number. *Annual Review of Psychology, 40,* 155–189.

Gallistel, C. R. (1990). *The organization of learning.* Cambridge, MA: Bradford Books/MIT Press.

Gelman, R., & Gallistel, C. R. (1978). *The child's understanding of number.* Cambridge, MA: Harvard University Press.

Gibbon, J. (1981). On the form and location of the psychometric bisection function for time. *Journal of Mathematical Psychology, 24,* 58–87.

Gibbon, J., & Church, R. M. (1990). Representation of time. *Cognition, 37,* 23–54.

Gibbon, J., & Church, R. M. (1981). Time left: Linear versus logarithmic subjective time. *Journal of Experimental Psychology: Animal Behavior Processes, 7*(2), 87–107.

Gibbon, J., Church, R. M., & Meck, W. H. (1984). Scalar timing in memory. In J. Gibbon & L. Allan (Ed.), *Timing and time perception* (pp. 52–77). New York: New York Academy of Sciences.

Herrnstein, R. J. (1990). Levels of stimulus control: A functional approach. *Cognition, 37,* 133–166.

Lea, S. E. G., & Dow, S. M. (1984). The integration of reinforcements over time. In J. Gibbon & L. Allan (Ed.), *Timing and time perception* (pp. 269–277). New York: Annals of the New York Academy of Sciences.

Matsuzawa, T. (1985). Use of numbers by a chimpanzee. *Nature, 315,* 57–59.

Mechner, F. (1958). Probability relations within response sequences under ratio reinforcement. *Journal of the Experimental Analysis of Behavior, 1,* 109–122.

Mechner, F., & Guevrekian, L. (1962). Effects of deprivation upon counting and timing in rats. *Journal of the Experimental Analysis of Behavior, 5,* 463–466.

Meck, W. H., & Church, R. M. (1983). A mode control model of counting and timing processes. *Journal of Experimental Psychology: Animal Behavior Processes, 9,* 320–334.

Meck, W. H., Church, R. M., & Gibbon, J. (1985). Temporal integration in duration and number discrimination. *Journal of Experimental Psychology: Animal Behavior Processes, 11,* 591–597.

Pepperberg, I. M. (1987). Evidence for conceptual quantitative abilities in the African grey parrot: Labeling of cardinal sets. *Ethology, 75,* 37–61.

Platt, J. R., & Johnson, D. M. (1971). Localization of position within a homogeneous behavior chain: Effects of error contingencies. *Learning and Motivation, 2,* 386–414.

Rilling, M. (1967). Number of responses as a stimulus in fixed interval and fixed ratio schedules. *Journal of Comparative and Physiological Psychology, 63,* 60–65.

Rilling, M., & McDiarmid, C. (1965). Signal detection in fixed ratio schedules. *Science, 148,* 526–527.

Rumbaugh, D. M., Savage-Rumbaugh, S., & Hegel, M. T. (1987). Summation in the chimpanzee (*Pan troglodytes*). *Journal of Experimental Psychology: Animal Behavior Processes, 13,* 107–115.

CHAPTER ELEVEN

Reflections on Number and Counting

Ernst von Glasersfeld
Scientific Reasoning Research Institute
University of Massachusetts
and
Institute of Behavioral Research
University of Georgia

When psychology, after the turn of the century, renamed itself the "study of behavior" and turned away from the profound thoughts of William James, Mark Baldwin, and other luminaries, it became unfashionable to go foraging in the fields of philosophy. This isolationism had profound consequences. It froze the psychological establishment in the conviction that their science had to fit the model of Newtonian physics and a simplified positivism was all that would be needed as epistemology. One manifestation of this development was the widely successful suppression of the difference between *empiricism* and *realism*. In the study of animal behavior, this led "hard-nosed empiricists" to the belief that the world as they conceived it (and therefore perceived it) was the only world there could be for human observers and animals alike. As a result, there seemed to be no need for the consideration that neither the perceptual environment (*Merkwelt*) nor the sphere of action (*Wirkwelt*) could possibly be the same for organisms endowed with different equipment for sensing and acting (von Uexküll, 1934). Epistemology was not something animal psychologists felt obliged to deal with.

The recent interest in "counting in animals," especially Davis and Pérusse's (1988) extensive discussion of the topic, suggests a change of direction. The very fact that there are references to Heyting, Peano, and Whitehead and Russell, cited because of their primary concern with the epistemology of number, shows that the interest goes beyond "observable behavior" and reaches far into the realm of hypothetical abstract concepts. At the same time, the discussion made it abundantly clear that among the contributors there are extremely diverse conceptions of number and what it means to "count." The ideas range from the inveterate *realist* view expressed in the statement "animals are said to exhibit 'numerical competence' when their behavior is controlled by a numerical dimension of the environment" (Nevin, 1988, p. 594) to a quotation of the *intuitionist* definition of the construction of number (Luchins & Luchins, 1988, p. 592). The divergence of attitudes and views thus suggests that the question, "What is truly numerical?", raised by Davis and Pérusse (1988, p. 609), would need an answer that can at least be tentatively agreed on, before an exchange of ideas about animals' competence in this area might become productive. In the hope of clearing the ground for some such working hypothesis, I examine the activities and operations with which the main number-related terms have recently been associated. Given that this analysis will inevitably be done from my point of view, I begin with a brief account of my own position concerning the notion of number.

PLURALITY AND NUMBER

In spite of "zero" and "one," it is safe to say that without the perception or conception of *plurality* there would be no concept of number. The first thing to ask, then, is how we come to have an idea of plurality. The raw material of this idea could not be anything but repetition. However, the mere taking place of repetition is obviously not yet the idea of plurality. The repetition has to be noticed, apprehended, or as Kant would say, to emphasize the active role of the subject, "apperceived."

To become able to recognize that some experiential thing or event turns up again and is the same, and thus repeated, is one of the indispensable conditions of conceptual development. Without the notion of sameness there would be no pluralities, no "sets," no numbers.

> Kinds and sameness of kind—what colossally useful *denkmittel* for finding our way among the many! The manyness might conceivably have been absolute. Experiences might have all been singulars, no one of them occurring twice. (James, 1907/1955, p. 119)

By specifying that sameness is a *Denkmittel,* William James put his finger on a crucial point. The composite German noun must be translated literally as "means of thinking." In the given context this accentuates the fact that "sameness" is not a property inherent in things, but a conceptual relation. No single thing can be "the same" as such. It has to be

11. REFLECTIONS ON NUMBER AND COUNTING 227

experienced more than once, and only a perceiving subject that has the ability to establish relations between its percepts, can judge that one occurrence is the repetition of another.[1]

Once such sameness judgments can be made, *pluralities* can be established. But these pluralities cannot be characterized except as "more than one," because mere repetition has neither coherence nor numerosity. The expression "abstracting an entity from perceived objects and indefinite repetition of an entity," quoted by Luchins and Luchins (1988, p. 592), does, as these authors noted, specify two notions that, for intuitionists, are "at the source of natural numbers." But they are not the whole of the intuitionist definition. The two notions define the generation of plurality. Number is for intuitionists, just as it was for Euclid, also "a unit of units." That is to say, it is a bounded plurality. Brouwer (1913), the founder of modern intuitionism, described this as

> the falling apart of moments of life into qualitatively different parts, to be reunited while remaining separated by time . . . the fundamental phenomenon of the human intellect, passing by abstracting from its emotional content into the fundamental phenomenon of mathematical thinking, the intuition of the bare two-oneness. This intuition of two-oneness, the basal intuition of mathematics, creates not only the numbers one and two, but also all finite ordinal numbers, inasmuch as one of the elements of the two-oneness may be thought of as a new two-oneness, which process may be repeated indefinitely. (p. 82)

Note that this definition of number, as far as I know the only one that attempts to define the concept in terms of the conceptual steps of the mind that generates it, is explicitly given as the definition of ordinal number, where the order is that of generation, and not yet the fixed sequence of standard number words.

In the human individual's ontogeny, a concept of number may arise in different ways. One that at least in our culture would be quite frequent, is the following: Let us say a little girl is holding up four fingers on one hand and has just counted them. At this point, there is the possibility that she will see both the four discrete fingers as counted units and the larger unit constituted by the bounded collection of fingers presented by her hand. If this happens, she will have acquired a perceptual embodiment of her count to "four." Such an embodiment may serve as a trigger for the *intuition* or, as Piaget would say, the *reflective abstraction* of a concept that could be described as "a unit of units." And it is precisely this

---

[1] The generation of the particular relation of *sameness* was investigated and given the name "assimilation" by Baldwin (1906–1911). The analysis of the process was refined by Piaget (1967), who made it into an integral part of his theory of cognitive equilibration.

dual nature of the conceptual entity that lays the foundation to the many more sophisticated abstractions that go to form the mathematician's conceptions of number. I want to emphasize that this is but one way for children to arrive at this fundamental concept. There are others, but all of them involve some form of reflection and abstraction.

On the level of human concepts, then, it is as a rule through the application of a number word sequence that the specific *numerosity* of a bounded plurality is established. The term *counting* in ordinary usage quite unequivocally refers to the ascertainment of a specific numerosity, and it remains to be seen, on the one hand, whether there are other ways of doing this and, on the other, whether one can find sufficient grounds to say that nonhuman animals count.

THE DEFINITION OF COUNTING

Brunschvicg, one of the great philosophers of mathematics at the turn of the century, quoted Lambert, a much earlier mathematician and astronomer, who in 1765 wrote in a letter to Kant: "The basis of science is not definition but all one has to know prior to creating a definition" (cited in Brunschvicg, 1912/1981, p. 484).

In the study of counting, this seems eminently relevant, regardless of whether we are concerned with human or nonhuman counters. Hence, I briefly lay out some of the things that, in my view, we do know about that particular activity.

For the adult human animal, counting is an everyday occurrence, almost as widespread and commonplace as the use of language. Human counting, in fact, relies on language in that it employs number words. Therefore, like language, it requires a form of reflective intelligence that goes somewhat beyond mere sensation and physical reaction.

In our everyday use of counting, it is not sufficient merely to know number words. An example should make this clear. Assume that someone counts the chairs in a room. The person runs through the standard number word sequence beginning from "one," comes to a stop, say, at "seven," but is then unable to decide whether the nine people who are coming to work in the room could all be seated. A witness would be astonished and might ask: "Didn't you count the chairs?"

I submit that this would be a perfectly normal reaction. We tacitly take it for granted that number words are used symbolically. In the given case this means that "seven" is not merely the tag assigned to the last available chair, but it also stands for the *whole count* and points to the fact that each of the preceding number words was also coordinated with an individual chair of the particular collection. Because the sequence of number words is used according to a conventionally fixed order, the last number word of the count also implies the bounded plurality with whose elements the sequence was coordinated and thus indicates its *numerosity*. We are expected to know that this implication is valid, not only in the case of chairs, no matter what items have been counted, and we are also expected to have arrived at the abstract notion that number words have a dual function: On the one hand they serve as standard terms in acts of counting, and on the other, each one of them can stand for a specific numerosity. This "standing for" or "meaning" something that is not the word itself, is what makes the use of words *symbolic*.

The statement that we tacitly take this symbolic use of the number words for granted, may easily be misunderstood. It may sound as though we knew more about counting than we actually do. As a rule, we are in no way aware of counting as involving a variety of mental operations; but we are thoroughly familiar with its results and the way they manifest themselves in actions. It is the production of these results that we tend to take for granted, even if we have never given a thought to some of the steps comprised in the activity that produces them.

One consequence of this lack of operational awareness[2] is that the term *counting* is frequently used for single segments of the procedure when there is no evidence whatever that the other component steps are executed. Conant's classical 1896 essay on counting (ca. 1900; see Conant, 1988) already illustrated this problem. Having cited the array of observations of "counting" in animals that prompted Lubbock (1885, p. 45) to say they seemed "very like a commencement of arithmetic," he commented that "many writers do not agree with the conclusions reached by Lubbock; maintaining that there is, in all such instances, a perception of greater or less quantity rather than any idea of number." Yet Conant went on to say: "But a careful consideration of the objections offered fails entirely to weaken the argument" (Conant, 1988, p. 425).

As in many more recent discussions, one is left with the impression that there is something ineluctable, something that has yet to be made explicit. In my view, this springs from the fact that the term *counting* ordinarily indicates a rather complex procedure but is then also used in other contexts where only a fraction of the component operations is manifested. Davis and Pérusse (1988) made a valiant effort to clarify the issue by speaking of "numerical competence" and subdividing this into several categories. Unfortunately the extensive discussion (pp. 580–602) tended to confound the suggested categories once more.

---

[2]As far as I know, it was Silvio Ceccato, in the 1940s, who coined the expression "operational awareness" (*consapevolezza operativa*) in the course of his seminal work on conceptual semantics (cf. Ceccato, 1964–1966).

Consequently, it may be worthwhile to approach the problem from a somewhat different angle by presenting a tentative operational analysis of the diverse situations to which the terms *counting* and *numerical competence* have been applied.

COUNTING TO DETERMINE A SPECIFIC NUMEROSITY

The example of the person coordinating the standard number word sequence and chairs in a room, illustrates what we have in mind when we use the word *counting* in a serious, practical context. In such situations, we require an explicit indication of the numerosity of a given collection of items. In terms of the distinction of "absolute" and "relative" numerosity (Thomas & Chase, 1980, 1988, p. 600), here it must be an absolute one.

In order to satisfy these conditions, the counter must be able to do the following things:

1. distinguish discrete *countable* items;
2. recognize a *plurality* of such items as the collection to be counted;
3. have an ordered sequence of individually different "tags," for example, the standard number word sequence;
4. coordinate the elements of this sequence in a *one-to-one* fashion with the items of the collection;
5. be aware of the *symbolic function* of the number words used—at least to the extent that, apart from serving as tag in the count, the last number word used can be interpreted as indicator of the counted collection's numerosity. (This last condition has been called "a sense of number" by some and "cardinality" by others.)

Vacuous Counting

Young children, once they are inducted into the use of words, can of course learn to say number words as well. Indeed, we have all heard parents ask toddlers to "count," at which the child often recites the first 10 or 20 number words of the standard sequence. But this "counting" should be considered vacuous because it does not involve the coordination of number words to countable items and, therefore, cannot establish any numerosity.

Numberless Counting

An equally rudimental but more useful procedure consists in the one-to-one coordination of the collection of items to be counted and a second, previously assembled collection of other items that are neither ordered nor individually significant (i.e., not number words or "numerons").[3] If the operation does not leave spares on either side, one can conclude that the numerosity of the two collections is the same. In this case, if the numerosity of the model collection is known before the exercise, it can of course be attributed to the new collection, but this is due exclusively to the successful one-to-one coordination and not to any numerical competence. At the most, one may say that this procedure leads to a *relative* or *comparative judgment of numerousness* (Kaufman, Lord, Reese, & Volkman, 1949; Thomas & Chase, 1980, 1988). Consequently, it can be considered evidence of the ability to carry out one-to-one coordination, but does not constitute an instantiation of the activity usually referred to as counting.

Mechanical Counting

One-to-one coordination of countable items and the terms of the standard number word sequence is relatively easy to mechanize and some gadgets we have come to depend on perform it very efficiently. They are commonly called *counters,* a usage I would consider *metaphorical.* It was therefore of interest to me to examine what operations these gadgets actually carry out.

The coordination of countable items and number words has three prerequisites: discrete unitary items, number words or numerals, and an act of coordination. It seems that merely *segmenting* something into discriminable units is not enough. Most trees generate discriminable strata in their trunks. When they are felled, these strata can be seen as rings that segment the duration of their life into years. But it takes an observer to coordinate the segments they create with number words. Hence, trees are not even metaphorically said to count.

In contrast, Gallistel (1988) gave as an example of a mechanism that counts "the small plunger-counter that door monitors in stores use." The gadget supplies only one of the three prerequisites: a mechanized number word sequence (it replaces the displayed numeral with its successor).

---

[3]Conant (1896) attributed this method to an 18th-century army official in Madagascar; my high school teacher, 60 years ago in Switzerland, explained it as the custom of shepherds in ancient Greece. Although it is an *intentional* activity to check a quantity and therefore requires some form of "conservation," I would not be surprised if it had been practiced in the Neanderthal.

The discrete countable items are generated by the door monitor who focuses on people walking into the store and coordinates each one with a push on the plunger. To call such a gadget a "counter" is, I submit, stretching even the metaphorical use of the term.

The wheel of a car that rolls over the road can be used to segment the distance it covers into countable units, provided we couple it to another wheel that has a tooth or indentation that touches a trigger once in every revolution (the countable unit). This trigger, then, does the door monitor's job of pushing a plunger that successively advances and displays the numerals of the preestablished standard sequence. This was the system of the first odometers that "counted" kilometers or miles. The fact that the mechanism has long been superseded by a more trouble-free electronic one, has not changed the principle: The odometer generates discrete segments and coordinates them with the standard number sequence, but it is always the user of the instrument who has to supply the notion of numerosity to the numeral the odometer displays.

An even simpler mechanical gadget was the water clock. It segmented an even flow of water into units by guiding it into a vessel. When the vessel was full, it tipped and emptied itself and, in doing so, rang a bell. Water clocks could be said to measure time and, consequently, might also be called counters.

I mention this last example, because the principle of *filling a preestablished unit* and thus giving rise to an act of counting would seem to be the principle of "number coding" that Thompson, Mayers, Robertson, and Patterson (1970) hypothesized as a possible counting mechanism in the cortex of cats. Basically this would be a neuron that has the property of firing whenever it has received a given number of single impulses from another neuron that generates them and, in doing so, creates the countable units. The collecting neuron, of course, has no notion of number but simply fires when it is "full."

As I mentioned (von Glasersfeld, 1988, p. 601) earlier, Dantzig (1930) suggested, in his famous book on the concept of number, that there might be internal "model collections," that is, templates of a certain numerosity, that could be used to match arrays or sequences of stimuli by one-to-one coordination. These templates would be equivalent to what Köhler (1940a, 1940b) called an "internal precept" and of which he explicitly said that their function should not be equated with human counting.

PATTERN RECOGNITION

The discussion generated by the term *subitizing* has been widespread and vigorous (cf. Davis & Pérusse, 1988) but, at least to my mind, not particularly enlightening. The trouble springs to a large extent from the fact that, from the very beginning, a numerical interpretation of an outcome tended to be confounded with the nonnumerical process of its production. The seeds of this entanglement were sown on the first page of the original paper by Kaufman et al. (1949),[4] the paper that launched the neologism. There, the authors said they are concerned with "the study of 'numerousness' by the direct method of reporting," which leaves their project somewhat obscure. Then they quoted Taves (1941), who spoke of "two 'mechanisms' for the discrimination of visual numerousness," and used the phrase "immediate and adequate perception of number." At the end, they concluded that their experiments supported Taves's inference that there are two mechanisms, one being counting, the other subitizing. It was this finding that prompted me to attempt an operational analysis of the phenomenon they described and labeled "subitizing" and that, in the conceptual development of children, may be quite independent of numerical concepts.

In the title of an earlier paper (von Glasersfeld, 1982), I deliberately mentioned "figural patterns" and reiterated throughout the text that the phenomenon I was analyzing and discussing had, as such, nothing to do with children's concepts of number or numerosity, but was due to the perceptual recognition of spatial or temporal configurations. Its connection to numbers, I stressed, was semantic.

Experimental psychologists who engage in studies of counting, be it in children or other animals, are as a rule competent counters and habitual users of numbers. At times this may lead them to confound an issue by introducing numerical terms into the explanation of a phenomenon that can be explained as easily or better in another way. A case in point is the phenomenon of *subitizing*. The phenomenon involves perceptual patterns that can be seen or heard as representations of numbers, so perceivers who are quite familiar with numbers will of course be able to see them as representing numbers, even if they have actually recognized them as spatial or temporal patterns. But, conceptually, it is one thing to recognize a spatial or temporal pattern as a pattern one knows and has associated with a certain name, and quite another to interpret the pattern as a collection of unitary items that constitute a certain numerosity.

With very young children, this is relevant because they sometimes have opportunities to associate a number name with familiar patterns of dots or other discrete items, before they have any conception of number or numerosity. I have seen this happen, but my evidence would, of course,

---

[4]The title of the paper by Kaufman et al. suggests they considered number a product of sensory discrimination rather than of conceptual operations; but given the date of publication, this might well have been a concession to the radical behaviorist fashion of the day. The results they presented, in any case, justify the hypothesis that one of the two mechanisms they separated has nothing to do with the concept of number.

be anecdotal. However, there is quite solid evidence that adults can use pattern recognition to discover the numerical value of conventional configurations that are partially hidden. This is very well known among bridge and poker players and is, in fact, the reason why they like to "play their hand close to the chest" (cf. the playing card example in von Glasersfeld, 1982).

When the term *subitizing* is introduced into the discussion of "numerical competence in animals," associations with standard number words seem out of the question. Hence, it may be in order to survey the diverse forms of pattern recognition that numerically competent observers might interpret as "apprehension of numerosity or number" and have been or could be included in the same category. I propose to do this according to the character of the patterns that are recognized.

Spatial Patterns

Arrays of discrete items in particular spatial arrangements are easily recognized if they fit a *canonical pattern* (i.e., patterns that are frequently perceived because they are more or less conventionally established or have been abstracted from recurrent "natural" experiences). If the canonical pattern is a *shape,* it will be triggered by the characteristic succession of movements, either of the eyes or of attention, within the visual field (Pritchard, Heron, & Hebb, 1960; Wertheimer, 1911; Zinchenko & Vergiles, 1972). This dynamic approach to pattern recognition was first used by Ceccato (1956) and later, independently, by Stallings (1977) in a system for the recognition of the basic components of Chinese characters.

If the pattern is constituted by a characteristic arrangement of specific items it may be recognized by the particular spatial relations that link the items. Premack (1975), in one of his imaginative experiments, showed that his chimpanzee, Sarah, was able to place two eyes, one nose, and one mouth appropriately into the blank outline of an ape's face. I would interpret this as evidence that she was operating with a canonical pattern.

Finally, if the pattern consists of a specific (small) collection of randomly arranged similar items, it could be recognized by a neuronal mechanism like the water clock (described earlier). One might suspect that the performance of infants who discriminate linear arrays of 1, 2, and 3 dots (Starkey & Cooper, 1980) is due to some such mechanism.

Note that especially in the last case outlined here, it is possible that recognition could be achieved not on the basis of the spatial pattern but by a count of the discrete items in an afterimage (Mandler & Shebo, 1982). However, I would suggest that this explanation would be plausible only when one is dealing with subjects who have clearly manifested at least one of the three components of the counting activity in situations where its use could be directly observed.[5]

Temporal Patterns

Temporal sequences of discrete but similar percepts can form patterns that may be characterized by the intervals between the individual percepts. Again there are at least two possibilities. The characteristic feature may be a stable sequence of intervals whose length constitutes a definite pattern, and in this case recognition will be due to the particular succession of durations, that is, their rhythm.[6]

Alternatively, the sequence may be characterized by the quantity of discriminable percepts, in which case recognition will once more be due to a mechanism similar to a water clock.

Functional Patterns

Another approach to problems that could be considered numerical but is yet unlike counting might be called *functional recognition*. Wertheimer, in his essay on the concept of number in native populations, described this with admirable clarity: "Tree trunks are being fetched. Corner posts or beams for building a house. One can count them. Or: one can go there, the picture of the house (the frame structure) in one's head, and fetch the trunks it requires" (1912, p. 109).

This is clearly another *figural* procedure. It differs from the one described earlier because the "picture in one's head" is characterized and determined not only by the sequence of movements in the perceptual construction of the pattern but also by the particular functions associated with the individual items that constitute the pattern. One can safely assume that this kind of functional recognition is practiced much more

---

[5]Here one could also include pattern recognition by a somewhat mysterious evolutionary process, such as the one primatologists hypothesize in the spontaneous recoiling of isolation-reared rhesus monkeys when they are faced with an adult's "threat posture" (Sackett, 1966). A similar phenomenon is the panic reaction to a hawk-shaped pattern that has been observed in domestic fowl (Tinbergen, 1948). Given the complexity of the patterns that are apparently recognized in these examples, it would not surprise me to learn that, say, tigers have a "canonic" pattern for four-legged prey, so that they can recognize the visual image "TTTTT" as an indication of more than one animal, without conceptualizing legs, counting up to 4, and a notion of "more than."

[6]Perception and re-presentation of rhythm by young children are the subject of excellent recent studies by Bamberger (1980, 1988). I do not know whether investigations of crickets, woodpeckers, fireflies, and so forth had similar success in separating the qualitative aspects of the discrete items from the temporal characteristics of rhythms; such research would seem to be relevant to the perception of "bounded pluralities."

frequently in everyday living than the recognition of canonical patterns of dots or lines.

Patterns and Numbers

Note that it is not only possible but extremely probable that in many living situations we use combinations of the procedures I listed as mutually independent methods of pattern recognition as well as combinations of these and such numerical knowledge as we have at the moment. In our work with children of the first two or three primary school grades (Steffe, von Glasersfeld, Richards, & Cobb, 1983; Steffe & Cobb, 1988) we have observed and recorded on videotape several situations where, in the course of solving an addition or subtraction problem, a child uses a "canonic" dot or finger pattern to record the intermediary result of a calculation. I would also claim that many adults, when faced with addition problems such as 48 + 12, derive the answer from a visualized *pattern,* such as stacks of 10, rather than from a memorized "number fact" or a counting operation.

Indeed, long before the word *subitizing* was invented, Wertheimer articulated the principle of figural solutions in unequivocal terms. His formulation was quoted by Luchins and Luchins (1988) and I repeat it here because it is directly pertinent to a point I want to make. "There are structures (*Gebilde*[7]) which, less abstract than our numbers, nevertheless serve analogous ends or can be used in place of numbers" (1912, p. 108).

If an observer, describing someone else's action, says "He is using A in the place of B," it does not tell us whether the actor knows B or is unaware of it. This would seem to be eminently relevant in the context of animal observation. If an animal behaves in ways that we could explain by any form of pattern recognition or subitizing, it would be unwarranted to attribute any notion of number or numerosity to that animal (cf. Davis & Pérusse, 1988, p. 609). On the other hand, it is clear that a subject who is familiar with numbers and counting can at any time interpret the results of an act of subitizing in terms of numbers. I presume this is similar to what Warren (1897, quoted in Mandler & Shebo, 1982) intended with the expression "inferential counting."

In defense of the original and painstaking work on subitizing by Taves (1941) and Kaufman et al. (1949), I want to counter Gallistel's argument about reaction times by the following consideration. Quoting one of the experiments by Mandler and Shebo (1982), he wrote, "It takes 30 msec longer to recognize twoness than to recognize oneness, 80 msec longer

---

[7]The word *Gebilde* could also be translated as "constructs" in order to emphasize the conceptual nature of the referent.

to recognize threeness than twoness, 200 msec longer to recognize fourness than threeness and from fourness on up 350 msec per item." He then argued that when (as in the experiment that furnished the quoted measurements) the arrays presented do not conform to a canonical pattern, "the reaction time function is what you would expect from a counting process: There is an increment in the reaction time for each additional item in the set to be counted" (Gallistel, 1988, p. 586). I find this unconvincing. If items are counted, each item of a presented array must be perceptually isolated, the appropriate number word of the standard sequence must be coordinated with it, and the last number word must be uttered. Given that the first four number words of our standard sequence all take very much the same time to utter, and perceptually isolating the last dot in an array of four cannot take very much longer than isolating the last dot in arrays of three, the increase of reaction time from one array to the next should remain very close to constant if one is counting (the additional work being simply the counting of one more item). In contrast, if one is perceiving and recognizing patterns, it is clear that the time requirement for both these operations will increase by some geometrical function of the area or the figural complexity of the pattern to be scanned, rather than linearly with the number of items displayed. I have no great faith in reaction time measurements as indicators of cognitive processing, but it seems to be the case that in this context they point in the direction of pattern recognition rather than counting.

ESTIMATING

Kaufman et al. (1949) presented their "operational definitions" of the two noncounting terms they have used,

> so that any person (otherwise able to do so) can carry out the operations and produce either subitizing or estimating as he chooses. As applied to the visual discrimination of numerousness in a briefly and simultaneously presented, randomly arranged field of dots: *estimating* is what occurs when the stimulus-number is greater than 6; *subitizing* is what occurs when the stimulus-number is less than 6. (p. 521)

The operations they spoke of are, quite clearly, operations the experimenter carries out, not those of the subject who manifests the observed phenomenon. They also noted, "Strictly speaking, however, there will be two kinds of numerousness. One kind will be associated with the process of subitizing. . . . The other kind will be associated with estimating" (p. 521). Strictly speaking, the two kinds of numerousness must be those experimenters distinguish according to the size of the array they happen to be presenting, not those that might be "visually discriminated" or "apprehended" by the subjects and cause them to manifest the discontinuity in the reaction times Kaufman and his colleagues found in their measurements.

In the tentative operational analyses I listed earlier, I was explicitly concerned with the operations of the acting subject. When we come to *estimating,* I am again primarily interested in finding out how the subject arrives at a result. However, the constraints the experimenter might impose to preclude that the subject reaches the result by means other than estimating, are interesting because they may indicate what operations the subject is not carrying out.

These constraints are two and they are combined in the presentation of the task: (a) The array contains too many discrete items (dots or whatever) to be recognized as a pattern, and (b) the presentation is too short to permit counting. In fact, estimation is a more or less educated form of guessing. In certain crafts (brick-laying, carpentry, etc.) it can be "educated" to a remarkably high degree, especially with regard to *continuous* quantity. But—and this is the salient point—even the best estimators remain aware of the fact that their judgments of numerosities are never as trustworthy as a proper count. Nor are they trusted as readily as the results of subitizing. When, for instance, you have recognized a pattern as a pattern of four, you have no doubt that you are right (even if later it turns out that you are wrong). You have seen it as a four, and in this regard seeing is believing.

In practice, estimating is mostly used for judgments of *comparative* or *relative,* not for *absolute numerosity.* Consequently, it rarely concerns numbers as such, and when it does, it is uncertain. I would add that it is presumed to be based on figural features such as spatial extension or density, rather than pattern, and the meaning of the word *estimating* would be betrayed if one used it in a case where one suspects actual counting.

How human and nonhuman subjects assess extension and density is no doubt an extremely difficult question to answer. For the purposes of the present discussion, it will suffice to say that judgments of *spatial extension* are likely to be based on the proprioception of movement carried out in the act of perceiving, whereas judgments of *density* would seem to derive from properties in the visual field, such as salience, brightness, or color. Neither the first nor the second are directly dependent on the discreteness of the items that form the estimated collection, so it follows that the transformation of the *figural* result of estimation into a *numerical* judgment requires the estimator's prior familiarity with the concepts of numerosity and number.

Another Perceptual Method

A somewhat related case, however, has been cited in the literature on counting in animals and therefore merits closer examination (King, 1988, p. 590). I am referring to the "middleness" experiments by Rohles and Devine (1966). Their study presented the astonishing finding that a chimpanzee was able to determine quite reliably the middle item of asymmetrical collections that contained more than a dozen discrete units.[8] Given that the spaces between the items in the presented linear arrays were deliberately unequal, the central item was not in the spatial middle of the array, nor could it be identified by a perception of symmetry, because the patterns on the two sides were in no way similar. Thus it looked very much as though the chimpanzee were counting or, in some other way, operating with numerosities.

It is a pity that the experiment did not include observation of the subject's eye movements. Consequently, one cannot rule out that the chimpanzee arrived at the correct result by a sophisticated use of proprioceptive data from scanning movements. To me this is an intriguing hypothesis and I am presenting it here in the hope that someone will test it.

To find the middle item in an array such as:

```
A B C    D    E F  G H  I
x x x    x    x x  x x  x
```

by following a procedure that simply keeps track of the distance covered by successive scanning movements in opposite directions, making sure that each movement ends on the last item before one has moved as far as in the preceding movement. Thus the first path will be from A to I, the second from I to B (but not to A), the third from B to H, and so on, until one arrives at E and finds that there is no more move shorter than the last one (from F to E) that would end on an item. A further perceptual help in this procedure would be to register the distance between the first and the second item on each move, because the second one will be the stopping point for the next return movement. The item where the last cross movement ended, then has to be the middle item.

I submit that with a little practice one becomes quite good at this, and chimpanzees, whose visual memory is certainly as good if not better than ours, should have no difficulty in learning the procedure—provided they think of it. I am not suggesting that they did or can. But I do want to

---

[8]A later study by the same authors is not so interesting in this context, because there the middle had to be found in an array of no more than 11 items, and (theoretically) this could be achieved by subitizing a pattern of 5 from both ends.

suggest that there may be many similarly *nonnumerical* methods that animals can use to solve problems that, to the human experimenter, seem approachable by a numerical procedure only. The often unconscious supposition that the way we tend to solve problems of a certain kind is the only way of solving them, is a widespread manifestation of *anthropocentrism* and seems almost unavoidable in cases where we have not yet thought of an alternative solution. However, when dealing with nonhuman organisms, whose experiential world may be very different from ours, it may be more profitable to try to observe everything the animal does, rather than focus only on such actions as we can relate to what we would do in similar circumstances. As Piaget always claimed, this "decentering" takes a long time to develop but is not unattainable. In my view, this decentering, is as relevant in animal psychology as it is an attempt to understand the thinking of young children.

CONCLUSIONS

In our studies of first and second graders' experiences with numbers, counting, and simple arithmetic (Steffe et al., 1983) we came to realize that problems a teacher presents in class can be interpreted by the children only in terms of their own concepts. Children, then, may produce a solution that makes good sense to them, but fails to be recognized as a solution by the teacher who has a different problem in mind. In such cases, much is lost if the teacher does not attempt to find out why and how the children operated. It was an opportunity to discover something about children's reasoning, something about why that particular solution made sense to them, and—more important from an educator's point of view—it was an opportunity to nurture the spark of satisfaction that may have accompanied finding a solution, even if it was not the solution the teacher wanted.

Although animal psychologists are rarely concerned about fostering motivation in their subjects (meat pellets or raisins are supposed to do this automatically), the question of how the animal subject sees and categorizes the experimental situation should be just as important here as it is with children. Only an extremely naive realist view of the world could lead one to assume that if an animal manages to discriminate numerically different arrays of items, it must be counting.

In the preceding pages I argued that there are several ways of operating that yield results that, to number-conscious human observers, seem compatible with the results they obtain by a count. It seems safe to predict that there are other ways we have not yet thought of and it would not be without interest to isolate and analyze as many as we can. I would even venture to say that the isolation and analysis of these possible ways of operating would be more interesting than arguments about whether or not a particular animal in a particular experiment was actually counting. Such arguments, given the idiosyncratic definitions of the term, are too reminiscent of those that raged around the question of whether or not the chimpanzees Washoe, Lana, Sarah, and the unfortunate Nim were using "language," at a time when there was no agreed definition of that term.

The phrase that Brunschvicg quoted from Lambert that I quoted earlier is eminently appropriate in this discussion. Today we know enough about the counting of children to formulate a viable definition of this activity in the human use of numbers. But when we are dealing with non-human organisms who, as far as we can tell, do not employ a standard number word sequence, we should find out much more about what these organisms do or cannot do, before we can meaningfully expand our human definition of counting to include some of their ways of operating.

REFERENCES

Baldwin, J. M. (1906–1911). *Thought and things or genetic logic: A study of the development and meaning of thought* (3 Vols.). New York: Macmillan.

Bamberger, J. (1980). Cognitive structuring in the apprehension and description of simple rhythms. *Archives de Psychologie, 48,* 171–199.

Bamberger, J. (1988). Les structurations cognitives de l'appréhension et de la notation de rhythmes simples. [Cognitive structuration in the perception and notation of simple rhythms]. In H. Sinclair (Ed.), *La production de notations chez le jeune enfant.* Paris: Presses Universitaires de France.

Brouwer, L. E. J. (1913). Intuitionism and formalism (Inaugural address at the University of Amsterdam, Oct. 14, 1912). *Bulletin of the American Mathematical Society, 20,* 81–96.

Brunschvicg, L. (1981). *Les étapes de la philosophie mathématique* [Stages in the philosophy of mathematics]. Paris: Blanchard. (Original work published 1912)

Ceccato, S. (1956). La machine qui pense et qui parle [The machine that thinks and speaks]. Proceedings of the First International Congress of Cybernetics. In *Actes du 1er Congres International de Cybernétique, Namur, 1956* (pp. 288–299). Paris: Gauthier Villars.

Ceccato, S. (1964–1966). *Un tecnico fra i filosofi* [A technician among philosophers]. (2 Vols.). Padua: Marsilio.

Conant, L. L. (1988). Counting. In J. R. Newman (Ed.), *The world of mathematics* (Vol. 1, pp. 423–432). Washington: Tempus Books. (Original work published ca. 1900)

Dantzig, T. (1930). *Number: The language of science.* London: Macmillan.

Davis, H., & Pérusse, R. (1988). Numerical competence in animals: Definitional issues, current evidence, and new research agenda. *Behavioral and Brain Sciences, 11*(4), 561–615.

Gallistel, C. R. (1988). Counting versus subitizing versus the sense of number. *Behavioral and Brain Sciences, 11*(4), 585–586.

James, W. (1955). *Pragmatism.* Cleveland/New York: Meridian. (Original work published 1907)

Kaufman, E. L., Lord, M. W., Reese, T. W., & Volkmann, J. (1949). The discrimination of visual number. *American Journal of Psychology, 62,* 489–525.

Köhler, O. (1940a). Vom Erlernen unbenannter Anzahlen bei Tauben [On pigeons learning unnamed numerosities]. *Veröffentlichungen der Reichsstelle für den Unterrichtsfilm, Nr. B 521,* 1–14.

Köhler, O. (1940b). Wellennsittiche erlernen unbenannte Anzahlen [Budgerigars are learning unnamed numerosities]. *Veröffentlichungen der Reichsstelle für den Unterrichtsfilm, Nr. B 523,* 1–14.

King, J. E. (1988). Number concepts in animals: A multidimensional array. *Behavioral and Brain Sciences, 11*(4), 590.

Lubbock, Sir John (1885). On the intelligence of the dog. *Nature, 33,* 45–46.

Luchins, A. S., & Luchins, E. H. (1988). Numbers and counting: Intuitionistic and gestalt psychological viewpoints. *Behavioral and Brain Sciences, 11*(4), 591–592.

Mandler, G., & Shebo, B. J. (1982). Subitizing: An analysis of its component processes. *Journal of Experimental Psychology: General, 11*(1), 1–22.

Nevin, J. A. (1988). Reinforcement schedules and "numerical competence." *Behavioral and Brain Sciences, 11*(1), 594–595.

Premack, D. (1975). Putting a face together. *Science, 188,* 228–236.

Pritchard, R. M., Heron, W., & Hebb, D. O. (1960). Visual perception approached by the method of stabilized images. *Canadian Journal of Psychology, 9,* 127–147.

Rohles, F. H., & Devine, J. V. (1966). Chimpanzee performance on a problem involving the concept of middleness. *Animal Behavior, 14,* 159–162.

Sackett, G. P. (1966). Monkeys reared in visual isolation with pictures as visual input: Evidence for an innate releasing mechanism. *Science, 154,* 1468–1472.

Stallings, W. W. (1977). Chinese character recognition. In K. S. Fu (Ed.), *Syntactic pattern recognition applications.* Berlin: Springer.

Starkey, P., & Cooper, R. G. (1980). Perception of numbers by human infants. *Science, 210,* 1033–1035.

Steffe, L. P., & Cobb, P. (1988). *Construction of arithmetical meanings and strategies.* New York: Springer.

Steffe, L. P., von Glasersfeld, E., Richards, J., & Cobb, P. (1983). *Children's counting types: Philosophy, theory, and application.* New York: Praeger.

Taves, E. H. (1941). Two mechanisms for the perception of visual numerousness. *Archives of Psychology, 37*(265), 1–47.

Thomas, R. K. (1988). To honor Davis and Pérusse and repeal their glossary of processes of numerical competence. *Behavioral and Brain Sciences, 11*(4), 600.

Thomas, R. K., & Chase, L. (1980). Relative numerousness judgments by squirrel monkeys. *Bulletin of the Psychonomic Society, 16,* 79–82.

Thompson, R. F., Mayers, K. S., Robertson, R. T., & Patterson, C. J. (1970). Number coding in association cortex of the cat. *Science, 168,* 271–273.

Tinbergen, N. (1948). Social releasers and experimental method required for their study. *Wilson Bulletin, 60,* 6–52.

von Glasersfeld, E. (1982). Subitizing: The role of figural patterns in the development of numerical concepts. *Archives de Psychologie, 50,* 191–218.

von Glasersfeld, E. (1988). Difficulties of demonstrating the possession of concepts. *Behavioral and Brain Sciences, 11*(4), 601–602.

von Uexküll, J. (with Georg Kriszat) (1934). *Streifzüge durch die Umwelten von Tieren und Menschen.* [Strolls through the environments of animals and humans]. Frankfurt am Main: Fischer.

Warren, H. C. (1897). Studies from the Princeton Psychological Laboratory, VI–VII. The reaction time of counting. *Psychological Review, 6,* 569–591.

Wertheimer, M. (1911). Experimentelle Studien über das Sehen von Bewegung. [Experimental studies on the perception of motion. Three theses on Gestalt theory]. *Zeitschrift für Psychologie, Bd. 60.*

Wertheimer, M. (1912). Ueber das Denken der Naturvölker, Zahlen und Zahlgebilde. [On the thinking of primitive tribes—numbers and numerical structures. Three theses on Gestalt theory.] *Zeitschrift für Psychologie, Bd. 61.*

Zinchenko, V. P., & Vergiles, N. Y. (1972). *Formation of visual images.* New York: C/B Consultants Bureau.

CHAPTER TWELVE

Chunking, Familiarity, and Serial Order in Counting

Wayne A. Wickelgren
Columbia University

At a Little League game awhile back, my 9-year-old son Peter told me, "Dad, this is a six Jason game!" By this he meant that there were six kids named Jason in the game—two on his team and four on the other. *How* one might recognize this is the topic of the present chapter. *Why* one might recognize it is another matter.

This chapter analyzes sequential counting and is not concerned with the discrimination of different numbers of simultaneously presented stimuli via subitizing mechanisms. Throughout this chapter, *counting* means *sequential counting*.

The chapter focuses on sequential counting for successive events where there is a substantial time interval between repetitions. When we sequentially count a set of simultaneous stimuli, we may use many of the same mechanisms required for sequential counting of successive events, but I have not considered this question in any detail. Thus, the emphasis here is quite different from that of Gelman and Gallistel (1978), for example, who concentrated on sequential counting of simultaneous stimuli.

This chapter considers sequential counting to require three basic competences: chunking, recognition of repetition, and the capacity to represent serial order. The serial order capacity corresponds to Gelman and Gallistel's stable-order principle. Neither repetition recognition nor chunking is present at all in Gelman and Gallistel's list of five basic principles that "govern and define counting," though the one-one principle has some similarity in function to repetition recognition and, there is a small functional overlap between chunking and Gelman and Gallistel's cardinality and abstraction principles.

It surprised me how much Gelman and Gallistel's model of counting relies on the simultaneous availability of the things to be counted and does not easily apply to counting successive events with long time intervals between repetitions. This is particularly noteworthy because Gelman and Gallistel argued against the claim that subitizing plays much of a role in the perception of even small numerosities of objects, and, like me, they are primarily concerned with sequential counting.

Gelman and Gallistel were also interested in children's understanding of number concepts, not just the ability to count, which I do not address at all. By contrast, I wish to make some progress toward more mechanistic, neural net, models of counting. Thus, although we share a focus on cognitive sequential counting by humans, there is very little overlap between this chapter and Gelman and Gallistel (1978).

The two major goals for this chapter are (a) to analyze the semantics of sequential counting by human beings in the sense of understanding what kinds of repeated events people count and (b) to analyze three component mechanisms that may play important roles in counting—chunking, repetition recognition, and serial ordering. The discussion of these mechanisms focuses primarily on a semantic or functional analysis of what these mechanisms accomplish and secondarily on making some progress toward a neural model of these mechanisms.

REPETITION, FAMILIARITY, AND NOVELTY

Semantics of Counting: The Role of Repetition Recognition

Counting Jasons is unusual in a baseball game, but counting balls, strikes, runs, and so forth is typical. How do you know when you have struck out? Umpires often use a mechanical or electronic counter to aid them, but batters usually know the count without the aid of auxiliary devices. Coding the number of strikes is paradigmatic of the human capacity for counting events.

Counting is not identical to coding duration, though there are times when we count the seconds. We can count events that occur with irregular time intervals between them, and experimental studies of counting often design counting tasks so that time duration is not too highly correlated with number of repetitions. Theoretical mechanisms for counting are likely to differ in some respects from mechanisms for encoding duration.

Counting is not limited to objects. We can count events of any type, including actions such as swinging a bat.

Counting is not limited to events that are identical or even very similar in their sensory qualities. The Jasons in Peter's game were all young boys, but they did not otherwise look much alike. The similarity was in their names, which I grant you is a physical sensory similarity. However, Peter did not recognize the similarity from hearing or seeing their names. He recognized the six Jason game by associative memory for the names from their disparate visual appearances. Finally, a strike in baseball is a disjunctive class of events that have no common physical component property that distinguishes them from balls, foul balls, hits, and so forth. There are called strikes that are over the plate and between the knees and the letters. There are swinging strikes where the batter misses the ball completely, foul ball strikes, and tipped balls that are caught by the catcher. In all of these latter cases it is irrelevant to classification as a strike whether the ball was in the strike zone.

However, although sensory, motor, and physical similarity are not necessary for counting, conceptual similarity of the events being counted is basic to counting and the recognition of repetition is a fundamental component of counting. What is repeated may be sensory or motor events that are nearly identical in some cases or merely possessing a single common sensory, motor, or cognitive attribute in other cases. We can count repetitions of bell rings, the number of objects found in a box, or the number of different outcomes in five throws of a die, but in all cases, there is an idea that classifies what is and is not to be counted. It is the ability to recognize that the representative of that idea has been activated repeatedly that lies at the base of our ability to count events.

I would like to dispense with a tricky problem by fiat. Although it can be argued that no two experiences are ever identical or identically encoded, it is clear that we often encode two events as equivalent in certain respects. I assume that in such cases there is some aspect of our representation of each event that is identical, though, to be sure, there are almost always other aspects of our encoding that are different on the two occasions. In my model, to recognize the repetition, the repeated part must at some point be the focus of attention, that is, be the complete thought active at that point in time. Perhaps the human mind is able to recognize and count repeated partial thoughts, but the model presented here only counts repeated complete thoughts.

Repetition, Recognition, and Recognition Memory

I remember that as a child I read *Treasure Island* seven times. My memory for that is probably no different from any other memory in how it is stored. But my ability to update the count each time I read it points out over how long a period of time one can count events and how extended and temporally complex the events may be. Like Dorothy following the yellow brick road to the Emerald City, I began at the beginning and, keeping my place via bookmarks and memory for what I had recently read, I continued to read, with frequent interruptions, to the end of the book, whereupon I activated the concept of "finishing *Treasure Island.*"

The first time I did this, I may well have simply encoded that I had read *Treasure Island,* not that I had read it once, though this is implicit. The second time I finished reading it and activated the "finishing *Treasure Island*" node, I somehow recognized that this had happened before. I recognized a repetition.

The recognition of repetition in this case is probably identical to what happens in recognition memory tests. We have a feeling of familiarity that is greater, on the average, for events that have been encoded by our minds in the past than for novel or unattended events. Thus, it would appear that explaining the ability to recognize repetition is identical to explaining recognition memory, though it is quite possible that this ability is mediated by more than a single mechanism and different mechanisms predominate in different cases, such as at short versus long time intervals between repetitions.

Familiarity/Novelty Detection Mechanisms

What are some possible mechanisms for recognizing repetitions in biological minds?

Immediate Repetition. One plausible algorithm for detecting immediate repetition in neural nets is to compute the sum of the absolute differences between the activation levels of each neuron at adjacent time periods as a fraction of the total activation level. Second-order immediate-change (velocity, difference, or derivative) detectors are quite common in real nervous systems—for example, motion, brightness changes, pitch transitions. Often some neurons respond to increases and others to decreases. The sum of the absolute differences would just be the sum of activity in both positive and negative change detectors. The total activation level (for the first-order neurons) must also be computed. Computing total activation in various subsets of neurons is relatively easy for neural nets. The primary problem in computing total activation is handling the wide range of totals, and real neural nets have obviously solved such range problems, as witnessed by our ability to encode brightness over an enormous range of light intensities.

A neural mechanism that encoded this sum of absolute changes over total activation is an immediate novelty detector, where immediate means over a time period on the order of a millisecond or few milliseconds. This mechanism computes novelty first, with repetition being signaled by very low novelty.

Clearly, this is not the sort of repetition-detection mechanism we need for cognitive counting, because the likely size of a time step for immediate repetition in the nervous system is on the order of a millisecond or a few milliseconds (basically neural communication delay times—axonal, synaptic, dendritic, and spike-generating delay times), and we can count events with seconds to years separating repetitions.

Long-term Repetition

Successor Thoughts. Perhaps we distinguish familiar versus novel thoughts on the basis of the different properties of the thoughts that follow the given thought. Although it may not always be true that recognition of familiarity implies some ability to recall associated ideas, it is often the case that when we recognize a familiar event, other ideas pop into our minds by association. Perhaps it is the immediacy and/or strength of activation of successor thoughts to a given thought that identifies that given thought as familiar versus novel.

For two reasons, I do not regard a mechanism based on the properties of successor thoughts to be as likely a basis for the identification of familiarity as a mechanism based on the properties of the given thought itself. First, such a mechanism is almost necessarily slower than one based on properties of the given thought. Second, because unfamiliar thoughts have familiar components and familiar components have associations, successor thoughts may well not provide much basis for distinguishing whether or not the prior thought was familiar or unfamiliar. In any case, I have no clear conception of how the properties of successor thoughts might distinguish familiar versus novel prior thoughts.

Chunking. It is likely that the mechanism for familiarity recognition is based on different properties of familiar versus novel thoughts during the activation of the given thought itself. The first time a particular thought is activated, chunking adds a new idea representative to the set of ideas and strengthens up associations from the constituents to the chunk and down associations from the chunk to its constituents. According to the basic hypothesis, a thought is judged familiar to the extent that it is already strongly associated to a single chunk idea. There are many possible specific mechanisms for this, three of which are discussed briefly here.

First, a thought may be judged to be novel if it triggers the chunking process, and familiar if it does not. The component of the chunking process that selects a new idea representative seems like an all or none event, whereas familiarity appears to come in various degrees. However, the strengthening of up and down associations could be a graded process that occurs to the greatest extent for novel or forgotten thoughts. Thus, it may be that some difference in the learning processes connected with chunking is the basis for familiarity recognition. However, because some property of the thought itself must trigger the chunking process, it seems more reasonable to make familiarity dependent on the earlier trigger property than on the later learning consequence.

One such trigger property might be complexity of coding or some other property that depends on whether or not there already is a single (chunk) idea to encode the current thought. That is, when an attention span of ideas is activated, retrieval may just automatically proceed to the highest, most economical level of coding, with chunks inhibiting their constituents after reaching full activation. If the final thought contains but a single chunk, then the thought is judged familiar. If the thought contains more than one idea, it is novel.

Familiar thoughts would have fewer active neurons than unfamiliar thoughts, and hence a control neuron or set of neurons that measures total activation would be a novelty detector. Novel thoughts would have more active neurons than familiar thoughts. Perhaps the recognition of novelty is based on activation of one or more neurons that innately encode the total activation of all neurons in some module or set of modules of the mind. Conscious recognition of novelty is limited to those modules of the mind of which we are conscious. These are presumed to be certain higher-order cognitive modules, and it is only in cognitive modules that I think chunking takes place (Wickelgren, 1979b).

However, it is not clear to me that thought activation automatically proceeds to the highest level in a single step. I think it is more likely that this requires several steps, with a set of constituents being activated in one step, and their chunk idea, if any, activated in the next step. It may be that the familiarity of an activated set of ideas is determinable from the activation of the set, before the chunk idea, if any, is activated. One way to determine the familiarity of a thought is as follows:

First, I assume that the human mind has multiple types of associations and can selectively enable and disable each type of idea and each type of association during different phases of thinking. Second, I assume that the mind has the ability to inhibit the currently active thought. Third, I assume that the mind has the ability to vary the maximum number of active ideas (neurons) in any particular phase of thinking.

To judge the familiarity of the active thought at time t, do the following for the transition to the next time step: (a) enable only the up associations, (b) set the threshold for activation of ideas sufficiently high so that only an idea with strong enabled input associations from all of the ideas in the prior thought could be strongly activated, and (c) temporarily inhibit the currently active thought. Then on or immediately after time $t + 1$, a control idea that judges the total activation at time $t + 1$ is a familiarity detector. The thought whose familiarity was to be judged was inhibited and only up associations were enabled from time t to $t + 1$, so only an idea with strong up associations from all of the ideas in the prior thought could be strongly activated at time $t + 1$. To the extent that any chunk idea is activated strongly, the familiarity idea, which judges total activation or maximum activation, will be strongly activated too.

This familiarity mechanism differs from the successor thought mechanism in limiting possible successors to ideas with strong up associations to the ideas in the current thought, whereas the successor thought mechanism permitted successor ideas with any of a variety of relations to the ideas in the prior thought, implicitly enabling all types of associations, including sequential associations.

Additional discriminative power is gained by setting activation thresholds high enough so that only a single idea receiving strong inputs from all of the ideas in the prior thought can be strongly activated.

Note that this familiarity mechanism uses the same sort of control idea measuring total activation as was used by the novelty mechanism discussed previously. First, this familiarity mechanism ought to give graded values of familiarity for repeated thoughts with different strengths of up associations to the chunk idea due to differences in forgetting or degree of learning. Second, the familiarity mechanism uses information gained from the very next time period after inhibition of the thought, which in a distributed neural net with attractor dynamics is generally many time steps prior to achieving a stable next thought. Thus, the familiarity mechanism acts fairly quickly. Third, this familiarity mechanism can be used to decide whether or not the mind should attempt to retrieve an existing chunk idea for this thought or select a new one. In my current neural modeling, this means enabling different types of links.

Short-term Repetition

Identification of repetition via a familiarity mechanism based on chunking seems highly appropriate for long-term memory counting, such as counting the number of times I read *Treasure Island*.

However, whether or not the current thought is already represented by a single chunk seems less appropriate for what could be called *short-term repetition,* namely, recognizing repetition, not over one's lifetime, but over some short period on the order of a few seconds. Such short-term repetition is still very long compared to what I referred to as *immediate repetition,* which was based on the cycle time of the nervous system—on the order of milliseconds. So the question is, do we need a third mechanism for repetition recognition over an interval of seconds?

It does not seem practical to develop a short-term repetition recognition mechanism out of the immediate repetition mechanism, because that mechanism uses second-order change neurons that monitor the activity of the first-order neurons and measures something approximating the derivative of the activation of the first-order neurons. In simpler language, such second-order neurons encode how much the first-order neurons changed activation from time step t_1 to time step t_2. To compute such differences between two nonadjacent activation states requires storing a sequence of activation states. This seems impractical over more than one or two intervening states and is highly implausible over a period as long as several seconds. Thus, we may confidently rule out extension of the immediate repetition mechanism to handle short-term repetition.

Consider an example of short-term repetition to see whether it can be handled by the long-term repetition mechanism. Imagine you are walking along the street and a nearby bell tower begins to ring out the hour of the day via a series of bongs. You count the bongs. How? You have heard these bongs many times before. You do not start the count where you left off last time, say at 347, and count 347, 348, 349, 350. No, you count 1, 2, 3, 4. Aha, four o'clock, you say to yourself.

There is difference in what is counted in short-term versus long-term recognition of repetition. Indeed, if you have been in this neighborhood on only a few occasions, you might also remember on how many prior occasions you have heard this bell chime before, perhaps on two prior occasions. You note that you have now heard this bell ring out the hour on three different occasions in your lifetime.

Nevertheless, the long-term recognition mechanism based on chunking can handle short-term recognition. Analysis of this problem deepens our appreciation of the cognitive complexities of repetition recognition and counting. The bell ring is familiar, so we could just recognize that and count lifetime frequency of our hearing this bell ring. But the context in which we hear the bell ring is different from past contexts (that is, the other ideas active in our mind are different), and we can form a chunk to represent the idea of the bell ring in this context the first time we hear the bell in the present context and thereafter count repetitions in just this context. The first time we hear the bell ring in the present context, we may just identify the bell ring as a familiar chunk in our long-term memory or we may follow this by encoding a new chunk of the bell ring in the present context. If we want to know what time it is, we must do the latter and count bell rings in this context alone, not over our whole lifetime.

SERIAL ORDER

One basic competence that underlies counting is knowing the order of the numbers, for example, being able to recite the sequence, "1, 2, 3, 4, 5, and so forth." Gelman and Gallistel (1978) referred to a slightly more general version of this competence by the term *stable order*. One might refer to this competence as rote counting, to distinguish it from other counting competences, such as repetition counting or object counting.

One might assume that representing the order of small positive numbers is more fundamental than, or at least different from, representing the order of items in other sequences, such as letters in written words or the alphabet or phonemes in spoken words. However, it seems more parsimonious to make the working assumption that the representation of ordered sets is the same in all of these cases.

I divide models of serial order into two large categories: nonhierarchical models that do not assume chunking to play any role in the coding of serial order and hierarchical chunking models that assume that long sequences are broken into subsequences. Subsequences are chunks, but these chunks have ordered sets of constituents.

Nonchunking Models

I do not regard nonchunking models of serial order as even remotely tenable, but brief consideration of them uncovers relevant types of associations and sets the problem in historical perspective.

Item-to-Item Association. In an associative memory, the simplest model for coding the order of the small positive numbers is to assume that there is a strong association from the 1 idea to the 2 idea, a strong association from the 2 idea to the 3 idea, and so forth.

However, a phenomenon often observed in children suggests this assumption is, at best, only a partial account of the associations involved in rote counting. The phenomenon purports that even children who are very accurate at counting from 1 to 10 often have considerable difficulty telling you the successor of some number between 2 and 9, for example, giving the answer to the question, "What comes after 4?" The prior number is not the only cue to the next number in the counting sequence.

Position Coding and Grouping. Essentially the same conclusion can be derived from a large number of studies concerned with determining the nature of the effective stimulus in serial list learning, namely, the prior item is not the only cue (Wickelgren, 1977, pp. 235–236; Young, 1968). People also use serial position as a cue.

Whether or not we use serial position as a cue for the next counting number is an interesting question to contemplate. At first, it may seem logically circular, like using an idea as a cue to itself. However, it is likely that serial position ideas are different from number ideas. Studies of serial position coding in short-term memory suggest that subjects often use only three different serial position ideas—beginning, middle, and end. However, they use these three position ideas in a cross-classification scheme at two different levels—beginning, middle, and end groups, and beginning, middle, and end positions-within-a-group (Wickelgren, 1964, 1967). In any case, it is very unlikely that we ordinarily use as many serial position ideas to code the order of items in a sequence as we have distinct counting number ideas.

Remote Associations. In addition to the prior item and serial position cues, we may also make some use of earlier, more remote, items as cues to the next item in a sequence. There is no definitive evidence that I know of to support this, but it is introspectively compelling to assign some cue value to remote prior items beyond their role in cuing serial position. So, it may be that $q\ r\ s\ t\ u$ is a better cue to the next letter v in the alphabet than u alone, because q, r, s, and t have remote associations to v.

Remote associations raise serious theoretical problems and have no definitive empirical support. There are more parsimonious explanations of the facilitatory effect of remote prior context including (a) more effective cuing of serial position, (b) inhibiting recall of the remote prior items as the successor item, and (c) cuing chunks representing sequences of items.

Thus, in naming the next letter of the alphabet, $q\ r\ s\ t\ u$ is better than u alone as a cue to v for any or all of the following reasons: (a) It more effectively cues that the desired letter is at a late-middle position in the second half of the alphabet (small effect). (b) It rules out q, r, s, and t as possible successors to u. (c) For all of us who know the alphabet song, it cues the chunks for *qrs* and *tuv* much better than does the single-letter cue u, with the chunk for *qrs* being associated to its successor chunk *tuv*. There are other chunking explanations as well.

Chunking Models

Although chunking in human learning and memory is not well understood, it is clear that humans chunk sets of ideas to obtain a more abstract representation of the entire chunk that is about as easy and simple to think with as the constituent ideas (Miller, 1956; Wickelgren, 1979a, 1979b). Chunking undoubtedly occurs for both ordered and unordered sets, but in the case of ordered sets (sequences), some additional property must be specified to encode the order information.

Note that a chunk for a sequence is a plan representative for the execution of that sequence. One aspect of planning in this model is to have a single abstract idea for a plan represent a sequence of constituent ideas.

As I see it, any viable model for serial order in long-term memory must assume that down associations from a chunk idea prime the chunk's unordered set of constituents. When these down associations are strong, this restricts activation to the constituents of the chunk out of the vast set of all possible ideas. In models of memory that assume chunking of sequences, the lion's share of the memory for a sequence is carried by the associations that pick out the unordered set of constituents, but the small remaining share that orders these constituents is essential, and there are a number of possibilities.

Gradient of Down Associations. It is possible that there is a gradient of down associative strength such that the strongest down association from a sequence chunk is to its first constituent, then its second, third, and so forth. Once a constituent node has been activated for some time, I assume that it gets inhibited, so the next most strongly activated constituent can become the most activated. Thus, when a sequence node is activated, it first activates its first constituent, then its second, third, and so forth. Grossberg (1978) employed a variant of this gradient model for sequence generation.

I have never liked this alternative, because I think it would be difficult to establish properly ordered and discriminable strength levels in learning that would correctly order the constituents in sequential retrieval. Many factors such as the discriminability, duration, and number of repetitions of constituents presented as stimuli ought to affect the strength of down associations, in addition to delay (remoteness).

It is interesting to consider what sort of learning process would establish this gradient of down associative strength. Superficially, it might seem that a simple contiguity conditioning process would do, because after a chunk is activated in production, the initial constituent of the chunk is the first to be activated, followed by the second, and so forth. However, it is logically circular to explain the ordering in production by a gradient in the strength of down associations and explain the gradient in the strength of down associations by the temporal ordering of the constituents in production. Some other learning process had to cause the gradient in down associative strength before the constituents of the sequence were first produced in the proper order.

The obvious candidate is the thought or perception that initially produced the sequence. But here the contiguity is probably reversed. The sequence of constituents is surely activated before the chunk representative is selected and activated, unless chunk representatives are selected in advance of their constituents and thus independently of them (assuming the usual forward direction of causality). If the first activation of a chunk follows activation of its constituents, if constituents of a sequence are activated in sequential order, and if down associations are strengthened in proportion to this contiguity, then it would seem that the strength gradient would be the reverse of what is required.

However, this is by no means necessary. For one thing, behavioral classical conditioning usually has an optimum interstimulus interval on the order of .5 s, not at 0. Perhaps all sequences that can be chunked are mentally activated within this half-second window yielding the desired gradient. A better argument offers that behavioral classical conditioning does not necessarily provide us with a direct window to observe the dynamics of associative change. We should not assume that all learning processes have the same temporal dependencies. Indeed, if up associations (uplinks) and down associations (downlinks) are different types of associations with different semantics and serving different functions, it would be quite reasonable for them to have different learning dynamics. Down associations might have a reversed strength gradient from up associations.

There is a more serious problem for the gradient model. In perception, constituents are not necessarily activated in the order of presentation. Longer, more intense, and, in general, more discriminable constituents may frequently be activated earlier, even if they are presented later. In a speed accuracy tradeoff study of speech recognition, Remington (1977) found that the medial vowel in a consonant-vowel-consonant (CVC) syllable is frequently recognized before the initial consonant. If perceptual recognition dynamics measures the dynamics of activation, this evidence suggests that neither recognizing the sequential order of stimuli, nor learning this order is dependent on the temporal order of activation. Although it is perfectly reasonable to permit the strength gradient model of order coding to use any functional relation between degree of learning and temporal contiguity of activation that achieves the functionally desired strength gradient, it is not obvious that the gradient model can survive order coding not being rigidly linked in some manner to the temporal order of constituent activation.

The Remington (1977) study also suggested that sequentially presented stimuli are often recognized in parallel. Although constituents with earlier stimulus onsets may increase in level of excitation earlier, the growth of excitation of all constituents of a short sequence may be heavily overlapping and the constituents that reach the threshold for activation first are often not those whose stimuli were presented first. After a constituent is activated, its activation may be maintained while other constituents are activated and each constituent may help to activate adjacent constituents by lateral sequential associative links prior to recruiting the chunk representative. Such dynamics of perception are difficult to square with the gradient model.

There is also another negative indicator for the model. In memory for sequences, it is almost always the middle items of the sequence that have the lowest probability of recall in response to a name for the sequence, suggesting, but not implying, that the down associations from the chunk idea to its middle constituents are weaker than those to either its initial or terminal constituents. The gradient model requires the association to the terminal constituent to be the weakest.

Jordan (1986) had an interesting discussion of the severe problems that the gradient model has with sequences that involve repeated elements, for example, *abcccd*. In such a sequence, the down association from the chunk to the *c* in the third, fourth, and fifth positions could easily be stronger than the association from the chunk to the *b* in the second position, causing *c* to be output before *b* erroneously. There are ways to overcome this problem. One way is to assume that the link strength to item i is greater than the sum of the link strengths to all items $j > i$. This requires a very large dynamic range of discriminable strengths for sequences of any appreciable length, but it might be viable for sequences of fewer than three or four elements. Another way is to recode sequences with repeated items into sequences of subsequences, where each sequence and subsequence has no repetition. This seems reasonable. Indeed, this latter approach may be needed to solve the problem of being able to activate a repeated item more than once, because an item is inhibited after it has been active for some time.

Despite many problems, a gradient of down associations from a sequence chunk is a possibility, especially if all long sequences are coded hierarchically such that no chunk has more than three or four constituents and no repeated constituents. The most attractive aspect of this alternative is the fact that only vertical associations are needed, up associations from constituents to the chunk and down associations from the chunk to the constituents. No lateral sequential associations are needed. All of the remaining models for serial order, with the possible exception of Jordan (1986), assume lateral associations somewhere, though not necessarily between the representatives of the elements of the sequence themselves.

Contingent Association. As Lashley (1951) pointed out, items like letters and numbers appear in many different sequences, so lateral associations from one such letter or number to another could hardly be relied on to carry the order information for each and every different sequence in which those items appeared. If the strongest associate to 1 is 2, that is fine for recalling the order of the cardinal numbers, but not so fine for recalling the successor digit of the first 1 in the value of pi, 3.14159.

One way to use lateral associations among constituents to mediate serial order is to make the strength of these lateral associations contingent on which chunk idea is active. Assume that an idea can have associations, not just to other ideas, but also to the associations between other pairs of ideas. Neurally, this is analogous to a link to a link, a synapse on a synapse.

In particular, assume that a sequence chunk idea sends down associations to the lateral sequential associations among its constituents. That is, chunks enable constituent lateral associations as well as constituent ideas. When the cardinal number chunk node is active, the sequential association from 1 to 2 is enabled by a down association from the cardinal number chunk, but when the pi chunk node is active, it is the sequential association from 1 to 4 that is enabled.

Rumelhart and McClelland (1986) implement such contingent associations by what they called *sigma-pi units*. The excitation of a normal unit (neuron) in most neural nets is the simple sum of the inputs received from all input links. The excitation of a sigma-pi unit in the present case would be the sum of the products of the input received from each lateral input link multiplied by the input from the most strongly activated down associative link from a chunk idea to that lateral link. Hence the name "sigma-pi"—the sigma is the summation operation applied to a bunch of terms that are obtained by multiplying two or more factors (the pi operation).

Sigma-pi units or associations to associations (links-to-links) are complex to implement in a neural net in both retrieval and learning, which is one strike against any contingent association model. A second strike against this particular contingent association model of sequence ordering is clear from the 3.14159 example, namely, the model does not discriminate different successors to repeated items at different positions of the same sequence, for example, to the 1 in 3.14159. Contingent associations can do anything that simple associations can do, so there is no question that an adequate contingent association model can be devised, if a simple association model can. The issue is whether we need to assume such more powerful and complex forms of associative memory. At this point, I think we do not.

Position Coding. Let the sequence *abc* be coded as *a1, b2, c3*. Let the sequence *cab* be coded as *c1, a2, b3*. When the constituents of a chunk are ordered, each constituent is itself a chunk with two constituents—a qualitative idea and a serial order idea. In retrieval of *abc,* the *abc* chunk excites its *a1, b2,* and *c3* constituents. Convergent excitation comes from a special serial order system that excites the "1" idea which, in turn, excites the "2" idea, and so forth. During the time the 1 idea is active, the 1 idea excites all of the chunks with a 1 constituent via up associations. This causes *a1* to be most strongly activated during the first time period of reciting *a1, b2, c3.* Via a down association from *a1* to *a, a* is the first item output in the sequence. Then via a strong innate or learned sequential association from 1 to 2, activation shifts to 2, which primes all of the chunks with 2 constituents, and so leads to the activation of *b2* and then *b,* and so forth.

Lashley's (1951) pioneering ideas concerning serial order are closer to position coding than to any of the other models discussed here, and MacKay (1987) developed this sort of serial order mechanism in considerable detail. MacKay's serial order mechanism is elaborated to handle serial order at each level of a multilevel hierarchy of constituent nodes. MacKay illustrated his model of sequence ordering by speech from the syntactic level of organization of phrases and words down to the phonetic levels of segments of words and features of segments. This is a necessary extension of any model of serial order, but it goes beyond the scope of this chapter.

Here I am concerned only with ordering the set of constituents of a single chunk. In position-dependent coding, a chunk first primes its unordered set of constituents, just as in every other model of serial order that assumes a role for chunking. In the position-dependent coding model the constituents of a chunk are somehow tagged for serial position, and a general serial order mechanism then activates the most strongly primed x1, followed by the most strongly primed x2, and so forth. The tags need not be labeled 1, 2, 3, and so forth, they could be labeled syntactic categories, such as "article," "number adjective," "size adjective," "color adjective," "noun," or "initial consonant group," "vowel group," as discussed in MacKay (1987). However, with respect to the control of serial order in production of a sequence, they might just as well be labeled 1, 2, 3. The labels become important when you relate these ordering ideas to other tasks. For example, if you want to cue a subject to produce one particular constituent of a sequence, then you need to use a label that communicates which constituent you want.

It is important to note that the position coding model of serial order has not eliminated sequential associations. To be sure, there are no sequential associations among the items of each and every learned sequence. But there are sequential associations among the generic position ideas. Consider the set of position ideas 1, 2, 3 used to order the sequence *a, b, c* by the coding *a1, b2, c3*. Neither the set of ideas *a b c*, nor the set *a1, b2, c3* needs to have sequential associations, but the set 1, 2, 3 does have to have sequential associations in order to impose its order on other sequences. There are two extant models of how this order is accomplished:

In the most obvious model, the command to produce a sequence activates the 1 idea. The 1 idea has its strongest sequential association to the 2 idea, the 2 idea has its strongest sequential association to the 3 idea, and so forth.

The less obvious model was suggested by Estes (1972) and was also favored by MacKay (1987). In this model the sequential associations are inhibitory, not excitatory. The 1 idea inhibits all the later position ideas. The 2 idea inhibits the 3, 4, and later ideas, and, in general, position idea i inhibits position idea j, if and only if $i < j$. This means that when the command to produce a sequence activates the set of position ideas, position 1 will win out initially. Then when the first item in the sequence has been produced, it must be inhibited, as must the position idea that activated it. Some active inhibition of a state to permit transition to a next state is undoubtedly a necessary component of any model of human thinking. In this case, once position 1 has been inhibited, position 2 will dominate all of the other serial position ideas and be maximally activated, and so forth.

Restricting the sequential links to one or more special sets of position ideas raises the issue of whether these sequential links among position ideas need to be learned. Clearly, they do not. The main advantage of the position model is that the sequential links can be innate, whether excitatory as in the first model or inhibitory as in the second model. In models that employ sequential associations among the items of each sequence to order the items in the sequence, it is necessary that these sequential associations be learned, that is, associations with modifiable strength.

Of course, while the lateral associations among the position ideas can be innate in the aforementioned position model, the association to each item must be learned, which completely negates any savings in number of learnable associations to encode a sequence compared to a model using lateral associations between adjacent items in the sequence.

However, there are other versions of the position model than the one I already described that make even the association from the position idea to each item an innate, rather than a learnable, association. The trick is to assume that each position idea is innately associated to a large set of representatives that can be used to represent every idea that ever appears in that position in a human's experience. This appears to be MacKay's (1987) model. When the sequences are syntactically structured sentences or phrases of words, it seems all right to assume that all noun ideas are innately associated to the noun position idea, all color adjectives are innately associated to the color adjective position idea, and so forth, though even here some might object to having separate representations for a word in each of the syntactic classes in which it can appear. However, this problem of multiple representations gets uglier and more ad hoc when, in order to learn arbitrary sequences of 5 words (the immediate memory span for words), we need to assume that every word has a separate representation for position 1, position 2, position 3, position 4, and position 5. And, of course, humans are capable of learning much longer random sequences of words.

But when humans learn long sequences, they typically employ grouping and other mnemonic devices that piggyback the new sequence on already learned sequences. I once studied grouping in short-term memory and concluded that in this context humans use just three position concepts—'beginning, middle, end'—but they can employ these in a cross-classification scheme consisting of beginning, middle, and end group with beginning, middle, and end position within a group (e.g., Wickelgren, 1964, 1967). It is important to note that the substitution errors tended to be from either the same group or the same position within a group, which indicates that the coding was more like a cross-classification scheme than a simple hierarchy.

Without a more detailed specification of the position model, it is not possible to conclude just how many different representations there would need to be for each idea to be able to distinctively encode its position in all of the different positions of all of the different kinds of sequences in which it could appear. At this point I see no reason to believe that the number of positional replications of each idea would be excessive.

My main reservation about position coding is rooted in the fact that I do not feel intuitively that humans use position information as much as prior item information in some of the sequences to which MacKay (1987) applied the position model.

Consider the issue of how we code the order of the phonemes in a word such as "strand," as discussed by MacKay (1987, p. 51). Assume that the subject has been shown the word strand and must now answer a yes–no question concerning a single constituent's relative position in the word. One could do an experiment to determine which of the following types of questions yields faster and more accurate retrieval (ideally by determining the speed–accuracy tradeoff functions for each): (a) Is the final nasal of the vowel group *n*? (b) Is the sound that follows /tra/ *n*?

There are lots of problems to designing this experiment so as to treat the issue of whether order is cued by position or prior items. For instance, I chose three prior phonemes, because that is how many prior phonemes are needed to cue a single context-sensitive allophone in the model to be described next, but it would be interesting to know how subjects would perform with a greater or lesser number of prior phonemes. Although I chose the same words MacKay (1987) used to describe his position coding model, I feel sure that MacKay would deny the validity of the first question as a test of his model on the very reasonable grounds that the position ideas of 'final nasal' and 'vowel group' are not going to be cued by the phrases "final nasal" and "vowel group," except possibly after some linguistic training, and not necessarily even then. There are lots of other problems. Yet I feel that pronouncing a few prior phonemes is a natural cue for the next phoneme in a word, and no position cue is equally natural.

In visual iconic memory, however, there is pretty strong evidence favoring position coding over prior item coding for serial order (Wickelgren & Whitman, 1970), and intuitively I feel that something like MacKay's syntactic categories plays an important role in ordering ideas at the level of concepts and words.

What is a plausible position coding model for the order of the small positive integer names that children learn when they learn rote counting or the order of the letter names in the alphabet that children probably learn in much the same way? The most plausible model is probably some multilevel grouping scheme, but I doubt that the groups are innate and invariant across all children. For numbers, no particular grouping seems terribly compelling. For the letters, *a b c* seems like a group, until we recall that in the alphabet song, the first group is *abcd*.

In principle, it is quite attractive to imagine that there is an innate representation of the basic number ideas from 1 to 7 or is it 1 to 3 or is it 1 to 9 or is it 1 2 3 many or . . . ? The evidence for three basic position cues provides some support for the 1 2 3 model, but counting gets to be pretty complex if you cannot have more than three different position ideas at each level of grouping. Here again, as in most other levels of coding below semantic memory, it seems intuitively wrong to think that humans are encoding order by means of some set of position ideas or banks of position-specific representations, except in the case of visual iconic memory, which is a very short-lasting nonassociative memory.

Context-Sensitive Coding. Although a single prior item often provides inadequate information to determine its successor in an unordered set, a sequence of two prior items is adequate in almost all cases. A plausible model to provide this information is provided by context-sensitive coding, which codes an idea context-sensitively with a different representation depending on prior and succeeding context (Wickelgren, 1969, 1972, 1979a). Although one might use more context, it appears to be sufficient to code a sequence of items by overlapping triples. Consider the sequence of five phonemes, /s t r u k/, composing the word "struck." In overlapping triple code, this is $_\#s_t, _st_r, _tr_u, _ru_k, _uk_\#$, where # means juncture, the boundary between two words. It is important to note that semantically, each phoneme triple means the particular allophone of the central phoneme that occurs in this left and right phonemic context. It does not mean a chunk of three phonemes. For this reason, I prefer to call these phoneme triples, *context-sensitive allophones,* or *c-s allophones,* to put the emphasis on the central element.

If accent is considered a distinctive feature of vowels, then to my knowledge, there are no ambiguities in ordering the unordered set of c-s allophones for any word in English. In any word, there is only one initial c-s allophone, that is an allophone of the type $_\#x_y$, and only one terminal allophone, that is of the type $_yx_\#$. Note that the context-sensitive coding model assumes a separate set of phoneme representatives for initial and terminal positions, as a simple position model might, but there must be more such representatives in the context-sensitive coding model, because each initial allophone is also conditioned by the second phoneme and each terminal allophone is conditioned by the next-to-last phoneme. The same internal allophone may appear in many different positions in different words, but it is likely that c-s coding requires considerably more phoneme representatives than any position coding model, which is a strike against it on grounds of representational efficiency, though the number of neurons required for c-s coding of phonemes is so tiny in comparison to the number of neurons in the nervous system that the force of this argument is much attenuated.

I believe it is useful vis-à-vis serial order to make a sharp distinction between creative behavior, such as the activation of sentences in thinking, and noncreative behavior, such as the activation of the phonetic constituents of words (Wickelgren, 1969). Lashley (1951) and MacKay (1987) do not. I think that c-s coding is simpler than position coding in learning and retrieval, and that it evolved earlier. As MacKay (1987) illustrated, position coding is a powerful device that can order words in novel utterances. Parrots can mimic words (noncreative serially ordered behavior), but they cannot create novel sentences. Perhaps it is precisely because they lack position coding capacity that birds' linguistic capacities fall short of ours.

I do not favor the version of position coding in which a set of units is innately dedicated to each syntactic category (position). Rather, I favor the version in which each concept or word has a syntactic category idea as a constituent. Then syntactic programs, coded via vertical and lateral associations among the syntactic categories, control the activation of concepts via up associations from syntactic category ideas to the concepts.

I believe these syntactic programs themselves rely on context-sensitive coding of syntactic categories, that is, that we use overlapping triples of syntactic categories such as $_{\#}\text{article}_{\text{size-adjective}}$, $_{\text{article}}\text{size-adjective}_{\text{noun}}$, $_{\text{size-adjective}}\text{noun}_{\#}$ to control the order of output of the words in a phrase such as "the large ball." I have a hunch that the evolutionary advance that permitted syntactically structured thinking was primarily in using order information stored in one system to control the order of activation in another system. Doubtless, MacKay and I will continue to agree to disagree.

Counting is not creative behavior, and, hence, I assume that our knowledge of the order of the small positive integers is context-sensitively coded in terms of overlapping triples of integers. Thus, the chunk 'count' has down associations to $_01_2$, $_12_3$, and so forth. The chunk for the ratio of the circumference of a circle to its diameter, pi, has down associations to $_{\#}3.$, $_{3\cdot}1$, $.1_4$, $_14_1$, and so forth. In context-sensitive coding the strong sequential associations are assumed to be between number triples, for example, $_23_4$ has a strong sequential association to $_34_5$. Although the digit sequences in some possible numbers would require richer context-sensitive coding than overlapping triples, such digit sequences are rare.

Jordan's Plan-State Model. Jordan (1986) proposed a neural net to code sequences that used three layers of neurons (input, hidden, and output), two classes of input neurons (plan neurons and state neurons), recurrent links among the state neurons, and backward (efference copy) links from the output neurons to the state neurons, in addition to the usual forward links from input neurons to hidden neurons and from hidden neurons to output neurons. Plan neurons encode the higher-level idea of the entire sequence, that is, an entire word, the alphabet, and so forth. The output neurons represent the lower-level constituents of the plan, that is, letters. State neurons represent serial order ideas, such as positions. Hidden neurons represent compounds of plans and states designed to permit nonlinear functions linking plans and states to constituents. Nonlinear functions are necessary to overcome the sort of associative interference, as pointed out by Lashley (1951).

A task Jordan gave to his network was to learn each of the six possible

12. CHUNKING, FAMILIARITY, AND SERIAL ORDER   265

sequences of three letters, *A, B, C,* that have no repeats: *ABC, ACB, BAC, BCA, CAB, CBA.*

One simple way to make such a network generate these sequences is to have one plan neuron for each of the six different sequences, one state neuron for each of the three positions in the sequence (first, second, third), one output neuron for each of the three letters, and one hidden neuron for each of the nine compound ideas: *A* first, *B* first, *C* first, *A* second, *B* second, *C* second, *A* third, *B* third, *C* third. It is necessary to assume an initial state for the state neurons, so we assume it to be that in which the first state neuron is active and no other state neurons. We assume that if the plan is to generate sequence *ABC,* then the *ABC* plan neuron is active and no others. Let the *ABC* plan neuron have a +1 link to the three hidden neurons: *A* first, *B* second, *C* third, and zero or negative links to the other six hidden neurons. Let the first state neuron have a +1 link to the three hidden neurons: *A* first, *B* first, *C* first, and zero or negative links to the other six hidden neurons. If the threshold for activating hidden neurons is 1.5, then only the *A* first hidden neuron will be activated to this input. Of course, we let the *A* first neuron have a +1 link to the *A* output neuron and zero or negative links to the other two output neurons, assume a threshold for all output neurons of .5, and then only the *A* neuron will be active as the first output of the *ABC* plan. Now if we assume that state one has a +1 link only to state 2, then state 2 will be active next, which, along with the maintained activation of the *ABC* plan neuron, activates only the *B* second hidden neuron, which activates the *B* output neuron, and so forth.

This simple model uses specific node coding, whereas Jordan uses distributed coding for inputs and outputs, and it makes no use of the backward links from outputs to state neurons, which Jordan admits are unnecessary to account for the basic ability to generate sequences. However, because distributed coding is simply a generalization of specific node coding, the previous simple, but high-interference, example makes it clear that Jordan's neural network has sufficient capability to overcome Lashley's serial order interference problem.

Jordan assumed the net learns the required link strengths by supervised learning, namely, the error (back) propagation learning algorithm of Rumelhart, Hinton, and Williams (1986) modified to handle recurrent nets. Personally, I think error propagation is too complex to be likely for real neural nets, but who can be sure of such matters? Anyhow, it enriches understanding to have contrasting theoretical alternatives, such as supervised versus unsupervised learning.

The specific example I described makes Jordan's model sound like the position coding model, but this is misleading. To be sure, Jordan's model can incorporate position coding ideas into the state and hidden neurons, but it can also incorporate prior context in a somewhat different way from context-sensitive coding via the backward efference copy links from output neurons to state neurons. The cue for the initial item of a sequence is the plan idea alone. The cue for the second item is the plan idea plus the initial item. The cue for the third item is the plan idea and the first two items of the sequence, and so on. This is called *efference copy* because some trace of the items remains active after they are output, until activation of some 'end' or 'juncture' idea terminates all activation associated with this sequence plan.

Jordan also used distributed coding instead of specific node coding. This can economize on the number of neurons required to code a set of ideas such as a set of sequences, positions, constituents, and so forth, but economizing on neurons in this way often leads to serious interference problems when learning many associations with the same net. I am quite enthusiastic about distributed coding, but I think that to avoid serious associative interference problems, it is necessary to implement chunking in distributed coding models. Hidden units in three-layer nets can perform many of the functions of chunking. Whether or not this is the optimal way to implement these functions remains to be seen.

COUNTING LONG-TERM REPETITIONS

A possible semantic model for counting long-term repetitions is as follows: When idea A is activated a second time, the familiarity idea is activated. If there are any associated memories for the number of prior activations of idea A, these are activated. Assume that the highest such number idea inhibits the others. Imagine that it is recalled that A has been activated on n prior occasions ($n > 1$), then what is activated is the digit triple $_{n-1}n_{n+1}$. Assume that simultaneous activation of the familiarity idea and the $_{n-1}n_{n+1}$ idea have previously been learned as the necessary and sufficient conditions for activation of $_{n}n + 1_{n+2}$, and so the count is incremented by one. Then A is chunked with the new count idea $_{n}n + 1_{n+2}$, so that the next time A is activated, it will activate the new chunk and its higher count $_{n}n + 1_{n+2}$.

When $n = 1$, the foregoing model works also, but I doubt that people always store a memory that an idea has been activated once, the first time it is activated. So one just adds to the model the proviso that if A is active, the familiarity idea is active, and no number idea is active, then it is $_{1}2_{3}$ that is activated and chunked with A. The next time A is activated $_{1}2_{3}$ will be activated along with the familiarity idea, which causes $_{2}3_{4}$ to be activated and chunked with A, and so forth.

This is no more than a brief sketch of a model for counting long-term repetitions. Many problems and surprises would occur in any attempt to implement this in a functioning neural net model or a more mathematical semantic model, but some model more or less along these lines could probably be developed.

It occurs to me now that we frequently know that we have experienced an idea on many prior occasions without recalling any specific count. We do not confuse such cases with cases where an item was experienced on one prior occasion and no explicit count of one occurrence was stored. Of course, there are ways to distinguish these cases, but until such matters are handled, the model's development is incomplete.

Nevertheless, it seems likely to me that any model of sequential counting of long-term repetitions in humans will involve submechanisms for chunking, familiarity recognition, and serial order.

REFERENCES

Estes, W. K. (1972). An associative basis for coding and organization in memory. In A. W. Melton & E. Martin (Eds.), *Coding processes in human memory* (pp. 161–190). Washington, DC: Winston.

Gelman, R., & Gallistel, C. R. (1978). *The child's understanding of number.* Cambridge, MA: Harvard University Press.

Grossberg, S. (1978). A theory of human memory: Self-organization and performance of sensory-motor codes, maps, and plans. *Progress in Theoretical Biology, 5,* 232–302.

Jordan, M. I. (1986). Serial order: A parallel distributed processing approach (Tech. Rep. No. ICS-8604). La Jolla, CA: University of California, Institute for Cognitive Science.

Lashley, K. S. (1951). The problem of serial order in behavior. In L. A. Jeffress (Ed.), *Cerebral mechanisms in behavior: The Hixon symposium* (pp. 122–130). New York: Wiley.

MacKay, D. G. (1987). *The organization of perception and action.* New York: Springer-Verlag.

Miller, G. A. (1956). The magical number seven, plus or minus two: Some limits on our capacity for processing information. *Psychological Review, 63,* 81–97.

Remington, R. (1977). Processing of phonemes in speech: A speed-accuracy study. *Journal of the Acoustical Society of America, 62,* 1279–1290.

Rumelhart, D. E., Hinton, G. E., & Williams, R. J. (1986). Learning internal representations by error propagation. In D. E. Rumelhart & J. L. McClelland (Eds.), *Parallel distributed processing* (Vol. 1, pp. 318–362). Cambridge, MA: MIT Press.

Rumelhart, D. E., & McClelland, J. L. (1986). *Parallel distributed processing* (Vol. 1). Cambridge, MA: MIT Press.

Wickelgren, W. A. (1964). Size of rehearsal group and short-term memory. *Journal of Experimental Psychology, 68,* 413–419.

Wickelgren, W. A. (1967). Rehearsal grouping and hierarchical organization of serial position cues in short-term memory. *Quarterly Journal of Experimental Psychology, 19,* 97–102.

Wickelgren, W. A. (1969). Context-sensitive coding, associative memory, and serial order in (speech) behavior. *Psychological Review, 76,* 1–15.

Wickelgren, W. A. (1972). Context-sensitive coding and serial vs. parallel processing in speech. In J. H. Gilbert (Ed.), *Speech and cortical functioning* (pp. 237–262). New York: Academic Press.

Wickelgren, W. A. (1977). *Learning and memory.* Englewood Cliffs, NJ: Prentice-Hall.
Wickelgren, W. A. (1979a). *Cognitive psychology.* Englewood Cliffs, NJ: Prentice-Hall.
Wickelgren, W. A. (1979b). Chunking and consolidation: A theoretical synthesis of semantic networks, configuring in conditioning, S-R versus cognitive learning, normal forgetting, the amnesic syndrome, and the hippocampal arousal system. *Psychological Review, 86,* 44–60.
Wickelgren, W. A., & Whitman, P. T. (1970). Visual very-short-term memory is nonassociative. *Journal of Experimental Psychology, 84,* 277–281.
Young, R. K. (1968). Serial learning. In T. R. Dixon & D. L. Horton (Eds.), *Verbal behavior and general behavior theory* (pp. 122–148). Englewood Cliffs, NJ: Prentice-Hall.

Author Index

A

Albert, M., 110, 116, *124*
Allport, D. A., 151, 153, *166*
Alptekin, S., 192, 196, 204, *208*
Amsel, A., 194, *207*
Anglin, J. M., 132, *144*
Antell, S. E., 159, *166*
Arndt, W., 20, *35*

B

Baldwin, J. M., 227, *241*
Balfour, D., 120, *124*
Bamberger, J., 156, *166*, 235, *241*
Barron, R. W., 110, *124*
Barry, K., 192, *208*
Bateson, G., 127, *144*
Beckwith, M., 157, *166*
Berg, R. F., 197, *207*
Berntson, G. G., 39, 40, 44, 45, 53, 57, 117, *124*, 138, *144*, 161, 162, *166*, 217, 219, 220, *222*
Bever, T. G., 87, *105*, 149, *168*
Bierens De Haan, J. A., 30, *35*
Blough, D. S., 162, *166*
Boakes, R., 16, *35*
Bolles, R. C., 194, *207*
Boysen, S. T., 39, 40, 42, 44, 45, 50, 53, 55, 56, *57, 58, 59,* 117, *124*, 138, *144*, 161, 162, *166*, 217, 219, 220, *222*
Brackbill, Y., 114, *124*
Bradford, S. A., 94, *105,* 110, 117, *124*
Brakke, K. E., 87, *106,* 243
Broadbent, H. A., 179, 183, *186,* 304
Brouwer, L. E. J., 227, *241*, 305
Brunschvicg, L., 183, 228, *241*
Burns, R. A., 164, 165, *167,* 175, 204, 205, *207*
Byrne, R. W., 65, 128, *147*

C

Cable, C., 63, *86*
Capaldi, E. J., 32, *35*, 39, *57*, 61, *85,* 112, 113, 114, 115, 116, 120, 121, 122, *124*, 134, 165, *167*, 192, 193, 194, 195, 196, 197, 199, 200, 201, 202, 204, 205, *207, 208,* 217, *222*
Ceccato, S., 229, 234, *241*
Chalmers, M., 49, *58,* 122, *125*

269

Chase, L., 143, *146,* 153, 155, *167,* 230, 231, *242*
Chi, M. T. H., 158, *167*
Church, R. M., 61, *85,* 111, *124,* 171, 172, 173, 174, 175, 176, 177, 178, 179, 183, 184, 185, *186, 187,* 215, 216, 218, 219, 220, *222, 223*
Cobb, P., 236, 240, *242*
Coburn, C. A., 16, *35, 37*
Colombo, M., 162, *167*
Colwill, R. M., 194, 198, *208*
Conant, L. L., 229, 231, *241*
Cook, R. G., 63, *86*
Cooper, R. G., 159, 160, *167, 168,* 234, *242*
Crane, D., 4, 5, 6, *35*
Crosby, T. N., 142, *146*
Curtis, L. E., 159, *169*

D

D'Amato, M. R., 132, 162, *167*
Dantzig, T., 232, *241*
Davis, H., 6, 10, 19, 22, 28, 30, 31, 32, 34, *35,* 39, 47, *57, 58,* 61, *85,* 94, 95, *105,* 109, 110, 112, 113, 114, 116, 117, 120, 121, 122, *124, 125,* 128, 129, 130, 135, 136, 137, 138, 139, 140, *144, 145,* 149, 150, 155, 160, 161, 162, 163, 164, 165, *167,* 171, 183, *186,* 193, 201, 202, 206, *208,* 216, 217, 219, *222,* 225, 226, 229, 232, 236, *241*
Delius, J. D., 63, *86,* 132, *146*
Devine, J. V., 239, *242*
Dewsbury, D. A., 128, *145*
Dickinson, A., 194, 198, *208,* 209
Dooley, G. B., 40, *58,* 130, 143, *145*
Dow, S. M., 221, *223*

E

Edwards, C. A., 63, *85*
Egeth, H., 135, *145,* 152, *167*
Emmerton, J., 22, *35*
Estes, W. K., 260, *267*

F

Fachinelli, C. C., 132, *146*
Feinberg, R. A., 195, 196, *208*

Fernandes, D. M., 172, *186,* 215, *222*
Ferster, C. B., 23, 24, 25, 29, *35,* 40, *58,* 93, *105,* 130, *145*
Fischel, W., 17, 19, *35*
Fitzgerald, H. E., 114, *124*
Folk, C. L., 135, *145,* 152, *167*
Fouts, D. H., 40, *58*
Fouts, R. S., 40, *58*
Fowler, H., 149, *167*
Fowlkes, D., 61, *86* 123, *125,* 130 134, 142, 143, *146,* 160, 162, 163, *169*
Fraisse, P., 156, *167*
Friedman, S., 159, *167*
Fuson, K. C., 39, 46, 47, *58,* 157, 158, 159, *167*

G

Gall, F. J., 9, 10, 32, 33, *35*
Gallis, P., 30, *35*
Gallistel, C. R., 6, 9, 11, 12, 27, 30, 32, *35, 36,* 39, 46, 47, *58,* 95, *105,* 112, 122, *125,* 136, 137, 138, 139, 140, 144, *145,* 149, 151, 152, 156, 157, 158, 159, 163, *167,* 171, 183, 184, 185, *186,* 193, 200, 206, *208,* 214, 217, 221, *222,* 231, 236, 237, *241,* 245, 246, 253, *267*
Gardner, B. T., 40, *58*
Gardner, R. A., 40, *58*
Garner, W. R., 151, *168*
Gelman, R., 39, 45, 46, 47, *58, 59,* 95, *105,* 112, 122, *125,* 136, 137, 138, 144, *145,* 149, 151, 152, 156, 157, 158, 159, 163, *167,* 171, 183, 184, 185, *186,* 193, 200, 206, *208,* 217, *222,* 245, 246, 253, *267*
Gibbon, J., 171, 176, 177, 185, *186,* 220, *222, 223*
Gill, T. V., 40, *58,* 130, 143, *145*
Gillan, D. J., 49, 50, *58,* 122, *125*
Gleitman, H., 83, *85*
Goodall, G., 198, *208*
Gordon, W. V., 205, *207*
Graham, H., 151, *167*
Greenfield, P., 87, *106*
Gregg, L. W., 152, *168*
Griffin, D. R., 111, *125,* 128, *145*
Groen, G. J., 45, 46, *58*
Grossberg, S., 255, *267*

AUTHOR INDEX

Grosslight, J. H., 196, *208*
Guevrekian, K., 172, *186*, 214, *223*

H

Haggbloom, S. J., 196, 197, *208*
Halford, G. S., 49, *58*
Hall, J. F., 132, *146*
Hammer, C. E., 24, 29, *35*, 130, *145*
Harlow, H. F., 89, *105*, 133, *145*
Hassenstein, B., 17, *36*
Hayes, K. J., 40, *58*
Heath, P. L., 131, *145*
Hebb, D. O., 234, *242*
Hediger, H. K. P., 13, *36*
Hegel, M., 39, *58*, 91, 94, *105*, 115, *125*, 130, 131, 139, *146*, 217, 220, *223*
Heron, W., 234, *242*
Herrmann, T., 116, *124*
Herrnstein, R. J., 63, *85*, *86*, 162, *167*, 217, *223*
Hicks, L. H., 39, *58*, 94, *105*, 129, 139, *145*
Hinton, G. E., 265, *267*
Hodos, W., 133, *145*
Holman, J. G., 198, *208*
Honig, W. K., 57, 61, 63, 79, 84, *85*, 149, *167*
Honigmann, H., 22, 30, *36* 129, *145*
Hopkins, W. D., 87, 90, 96, 102, 103, *105*, *106*
Hull, C. L., 194, *208*
Hulse, S. H., 149, *167*
Hunter, W. S., 15, *36*
Hurwitz, H. M. B., 28, 114, 116, *124*, 183, *186*

I

Ichihara, S., 152, 153, *168*
Ifrah, G., 138, *145*

J

James, W., 226, *241*
Jaroff, L., 115, *125*
Jensen, E. M., 151, *168*, 282
Jobe, J. B., 195, 196, *208*
Johnson, D. M., 214, 215, 220, *223*

Johnson, M., 113, *125*
Jordan, M. I., 257, 264, 265, 266, *267*

K

Kaufman, E. L., 134, 135, *145*, 149, 151, *168*, 231, 233, 236, 237, 238, *242*
Keating, D. P., 159, *166*
Keller, F. S., 132, *145*
Kendler, H. H., 131, *145*
Kendler, T. S., 131, *145*
Kikuchi, T., 152, 153, *168*
Kimble, G. A., 132, *145*
King, J. E., 239, *242*
Kingma, J., 49, *58*
Kinnaman, A. T., 11, 14, 15, 30, 33, *36*
Klahr, D., 150, 151, 152, 153, 157, 158, 160, *167*, *168*
Koehler, O., 7, 17, 18, 20, 21, 22, 33, *36*, 47, *58*, 112, *125*, 159, *168*
Köhler, W., 19, *36*, 232, *242*
Konorski, J., 194, *208*
Kuhn, T. S., 4, 33, *36*
Kwak, H-W., 135, *145*, 152, *167*

L

Landauer, T. K., 152, *168*
Lashley, K. S., 162, *168*, 258, 259, 263, 264, 265, *267*
Laties, V., 172, *186*
Lea, S. E. G., 63, *86*, 221, *223*
Leader, L. G., 172, *187*
Levine, M., 133, *145*
Lewandowski, A. G., 12
Lindsay, W. L., 9, 32, *36*
Littlejohn, R. L., 195, 196, *208*
Lombardi, C. M., 132, *146*
Lord, M. W., 134, 135, *145*, 149, 151, *168*, 231, 233, 236, 237, 238, *241*
Lorden, R. B., 39, 95, 128, *146*
Loveland, D. H., 63, *85*
Lubbock, Sir John, 229, *242*
Luchins, A. S., 226, 227, 236, *242*
Luchins, E. H., 226, 227, 236, *242*
Lynch, D., 195, *208*

M

MacFadden, L., 116, *124*, 341
MacKay, D. G., 259, 260, 261, 262, 263, 264, *267*

MacKenzie, K., 110, *124*
Mackintosh, N. J., 194, 195, 198, *208, 209*
Macphail, E. M., 163, *168*
Mahoney, M. J., 128, *146*
Mandler, G., 122, *125*, 135, 139, *146*, 154, 155, 158, 161, 162, *168*, 200, *209*, 234, 236, *242*
Marioq, A. V., 172, 175, 176, 185, *186*
Matlin, M., 140, *146*
Matsuzawa, T., 39, 40, *58,* 94, *105,* 160, 161, *168*, 217, *223*
Mayers, K. S., 156, *169*, 232, *242*
McClelland, J. L., 258, *267*
McDiarmid, C., 26, 215, *223*
McDonald, K., 87, *106*
McGonigle, B. O., 49, *58,* 113, 122, *125*
McIntire, R. W., 28, *35*
McPherson, J., 158, *169*
Mechner, F., 172, *186*, 214, 215, 220, *223*
Meck, W. H., 61, *85,* 111, *124*, 171, 172, 173, 174, 176, 177, 178, 183, 184, 185, *186*, 216, 218, 219, 220, *222, 223*
Medin, D. C., 207, *209*
Mellgren, R. L., 195, 196, *208*
Memmott, J., 22, 28, 31, *35*, 94, *105*, 110, 112, 114, 116, 121, *124*, 129, 130, *145*, 160, 163, *167*, 183, *186*, 193, 206, *208*
Menzel, E., 40, *58,* 90, *105*
Messenger, J. F., 154, *168*
Meyer, B., 27, *37*
Miller, D. J., 28, 32, *35*, 39, 57, 61, *85*, 112, 113, 114, 120, 121, 122, *124*, 164, 165, *167*, 202, *208, 222*
Miller, G. A., 142, *146*, 191, 192, 194, 196, 197, 199, 200, 201, 204, 205, *209*, 217, 255, *267*
Mitchell, R. W., 128, *146*
Morgan, C. L., 11, *36*, 128, *146*
Morris, M. D., 197, *207*
Morrison, S., 110, *124*
Mountjoy, P. T., 12, *36*

N

Nawrocki, T. M., 194, *208*
Neisser, U., 154, *168*

Nelson, K. E., 53, *58*
Nevin, J. A., 132, *146*, 226, *242*
Newell, A., 152, *168*
Nissen, C. H., 40, *58*
Noble, L. M., 133, *147*

O

O'Donnell, J. M., 9, 14, *36*
Oden, D., 40, *58*
Olshavsky, R. W., 152, *168*
Overmier, J. B., 194, 198, *209*
Oyama, T., 152, 153, *168*

P

Parkman, J. M., 45, *58*
Pate, J., 89, 91, *105,* 115, *125*
Patterson, C. J., 156, *169*, 232, *242*
Pearce, J. M., 149, *168*
Pepperberg, I. M., 29, *36*, 39, *58,* 113, 117, *125*, 160, 161, *168*, 217, *223*
Pérusse, R., 6, 10, 19, 22, 28, 30, 31, 32, 34, *35*, 39, 47, *58,* 61, *85*, 93, 95, *105*, 109, 110, 112, 113, 121, 122, *124*, 130, 135, 136, 137, 138, 139, 140, *145*, 150, 155, 160, 161, 162, 163, 164, 165, *167*, 171, *186*, 201, 202, 206, *208*, 219, *222*, 225, 226, 229, 232, 236, *241*
Peterson, G. B., 198, *209*
Piaget, J., 157, *168*, 227
Platt, J. R., 214, 215, 220, *223*
Popper, K., 4, *36*
Porter, J. P., 14, 15, 30, *36*
Povinelli, D. J., 53, *58*
Premack, D., 40, *58, 59,* 131, *146*, 234, *242*
Pritchard, R. M., 234, *242*
Prokasy, W. F., 132, *146*

R

Radlow, R., 196, *208*
Raskin, L. S., 55, 56, 57
Reese, E. P., 151, *168, 241*
Reese, T. W., 134, 135, *145*, 149, 151, *168*, 231, 233, 236, 237, 238

Remington, R., 256, 257, *267*
Rescorla, R. A., 194, 198, *208*
Resnick, L. B., 46, *58*
Restle, F., 157, *166*
Richards, J., 236, 240, *242*
Richards, R. J., 4, 10, 11, 34, *35*
Richardson, W. K., 90, *105*
Rigby, R. L., 195, 196, *208*
Rilling, M. E., 26, *36*, 39, 112, 113, 114, 215, *223*
Rivera, J. J., 63, *86*
Roberts, S., 172, 176, 177, *186*, *187*
Robertson, R. T., 156, *169*, 232, *242*
Rohles, F. H., 239, *242*
Roitblat, H. L., 87, *105*, 149, *168*
Romanes, G. J., 3, 10, *36*, 127, *146*
Rosch, E., 136, 140, 141, *146*
Rose, A. P., 159, *168*
Rosenthal, R., 12, 13, *36, 37*, 117, 118, 119, 120, 121, *125*
Rubert, E., 87, *106*
Rumbaugh, D. M., 39, 40, 42, *58*, 87, 89, 90, 91, 93, 94, 96, 101, 102, 103, *105*, *106*, 115, *125*, 130, 131, 139, *146*, 149, 162, *168*, 217, 220, *223*
Rumelhart, D. E., 258, 265, *267*
Runfeldt, S. A., 96, *105*

S

Sackett, G. P., 235, *242*
Salman, D. H., 22, 30, 31, *37*, 112, *125*, 129, *146*
Salmon, D. P., 162, 164, *167*
Saltzman, I. J., 151, *168*
Samuelson, F., 8, *37*
Sanders, R. E., 164, *167*, 204, *207*
Sands, S. F., 63, *86*
Savage-Rumbaugh, E. S., 39, 40, 42, *58*, *59*, 87, 90, 91, 94, 96, *105*, *106*, 115, *125*, 130, 131, 139, *146*, 217, 220, *223*
Schiemann, K., 20, 31, *37*
Schlosberg, H., 151, 154, *169*
Schoenfeld, W. N., 40, *58*, 132, *145*
Seboek, T. A., 13, *37*, 117, 120, *125*
Seligman, M. E. P., 27, *37*
Sevcik, R. A., 87, *106*
Shattuck, D., 116, *124*, *125*

Shebo, B. J., 122, *125*, 135, 139, *146*, 154, 155, 158, 161, 162, *168*, 200, *209*, 234, 236, *242*
Silverman, I. W., 159, *168*
Simon, H. A., 152, *168*
Sjoberg, K., 158, *169*
Skinner, B. F., 8, 22, 23, 25, *35, 37*
Smith, E. E., 207, *209*
Spence, K. W., 17, 30, *37*, 194, *209*
Sperling, G., 153, *168*
Spivey, J. E., 196, *208*
Stallings, W. W., 234, *242*
Starkey, P., 45, *59*, 159, *168*, 234, *242*
Steffe, L. P., 236, 240, *242*
Steirn, J., 87, *105*, 149, 162, 168, *169*
Stewart, K. E., 61, 62, 79, 84, *85*
Strauss, M. S., 159, *169*
Svenson, O., 158, 169
Swets, J. A., 25, *37*

T

Taves, E. H., 150, *169*, 233, 236, *242*
Terrace, H. S., 73, *86*, 87, *105*, 149, *168*
Terrell, D. F., 134, 135, 140, 141, 142, *146*
Thomas, R. K., 39, 61, *86*, 95, 123, *125*, 128, 130, 131, 132, 133, 134, 135, 137, 140, 141, 142, 143, *146*, *147*, 160, 162, 163, *169*, 230, 231, *241*
Thompson, N. S., 128, *146*
Thompson, R. F., 156, *169*, 232, *242*
Thompson, R. K. R., 40, *58*
Thorndike, E. L., 3, 37
Thorpe, W. H., 164, *169*
Tinbergen, N., 18, *37*, 235, *242*
Tolman, E. C., 194, *209*
Trapold, M. A., 194, 198, *209*
Tucker, M. F., 156, 159, *167*

V

van Cantfort, T., 40, *58*
van Kempen, H., 155, *169*
Vaughan, W., Jr., 63, *86*
Vergiles, N. Y., 234, *243*
Verry, D. R., 28, *35*, 113, *124*, 194, 196, 197, *208*

Vickery, J. D., 61, *86*, 123, *125*, 130, 134, 142, 143, *146*, 160, 161, 163, *169*
Volkmann, J., 134, 135, *145*, 149, 151, *168*, 231, 233, 236, 237, 238, *241*
von Glasersfeld, E., 122, *125*, 140, *147*, 155, 156, 157, 161, 164, *169*, 232, 233, 234, 236, 240, *242*
von Uexkull, J., 225, *242*

W

Wallace, J. G., 151, 152, 153, 157, *168*
Warren, H. C., 154, *169*, 236, *242*
Washburn, D. A., 90, 96, 102, 103, *105, 106*, 127, 128
Washburn M. F., 128, *147*
Wasserman, E. A., 63, *86*
Watson, J. B., 15, 16, *37*
Webster, J. B., 172, *187*
Wertheimer, M., 234, 235, 236, *243*
Wesley, F., 22, 31, *37*, 61, *86*, 112, *125*, 129, 130, *147*
Whiten, A., 128, *147*
Whitman, P. T., 262, *268*
Wickelgren, W. A., 250, 254, 255, 261, 262, 263, *267, 268*
Wijhuizen, G., 155, *169*
Wilkie, D. M., 172, *187*
Williams, R. J., 265, *267*
Wolters, G., 155, *169*

Subject Index

A

Absolute numerousness judgments, 142, 143, 144
Absolute numerosity, 62, 70, 73, 80, 81, 82, 84, 109, 238
Absolute prototype matching, 136
Abstraction principle, 95, 137, 245
Addition, 44, 45, 46, 213, 214, 219, 221
Addition algorithms, 45, 46
Animal thinking, 111
Arabic numerals, 41, 42, 43, 44, 45, 93, 94, 96
Arithmetic reasoning, 211, 212, 219, 220, 221
Associative learning, 194, 195, 197, 198, 207, 253, 254, 255, 256, 257, 258
Autocontingencies, 116

B

Behaviorism, 8, 9, 15, 23, 24

C

Canonical pattern, 154, 234
Cardinal principle, 95, 137, 183, 184, 218, 219, 222, 245
Cardinality, 95, 230
Carrier stimuli, 82
Chunking, 245, 246,
Chunking models, 255, 256, 257
Chunking lists, 192
Chunking series, 192, 207
Classical concepts, 207
Classical conditioning, 194, 198, 199
Clever Hans, 12, 13, 18, 111, 117, 118, 119, 120, 128, 129, 149, 201
Cognitive revolution, 8, 23
Comparative psychology, 4, 11, 13, 14, 15, 34
Concept of number, 135, 136
Conditioned suppression, 27, 28
Connectionist models, 179, 180, 181, 182, 183, 184, 185
"Counting all" strategy, 46
Counting, definition of, 12, 30, 32, 61, 62, 95, 111, 112, 131, 136, 137, 144, 183, 184, 185, 192, 193, 217, 218, 221, 222, 226, 228, 230, 231, 232
"Counting on" strategy, 46

D

Devaluation studies, 198
Differential outcome studies, 198

275

Discrimination ratio, 67, 68, 69, 70, 71, 72, 73, 74, 76

E

Entry-level counting, 101
Enumeration, 95, 203, 204, 205, 206
Estimating, 134, 136, 137, 141, 144, 149, 237, 238
Ethology, 4, 17, 18, 23, 34
Extinction, 196

F

Figural patterns, 155, 156, 233
Fractions, 52
Functional counting, 44
Functional recognition, 235

H

History of science, 4, 225

I

Instrumental learning, 191, 192, 194
Internal (timing) pacemaker, 175, 176, 177, 178, 179, 218
Invisible counting, 3

L

Learning set, 89, 133, 134
Learning theory, 4, 15, 16, 23, 24, 46, 192, 193

M

Matching-to-sample, 21
Mental category, 212
Mental primitive, 222
Methamphetamine, 175, 176, 179
Mind, 10, 211, 212
Morgan's canon, 11, 128
Motor tagging, 46, 47, 48, 49
Multiplication, 213, 214

N

Number comprehension, 42, 43
Number concept, 211, 212, 213, 219, 221, 222, 226, 227
Number constancy, 83
Number, definition of, 227
Number discrimination, 62, 65, 66, 68, 72, 78, 139, 172, 174
Number tags, 112, 149, 159, 183, 203, 204, 205, 206, 207
Numerical categories, 211, 212, 213, 214, 221
Numerical reference, 211
Numerical representations, 46, 55, 56, 57
Numerons, 112, 159, 217, 218, 231
Numerosity, 61, 62, 64, 82, 211, 214, 215, 216, 228, 229
Numerosity discrimination, 62, 63, 68, 79, 82
Numerosity judgments, 150

O

One-to-one correspondence, 40, 41
One-to-one principle, 137, 183, 184, 185, 206, 217, 222, 245
Order-irrelevance principle, 96, 137, 183, 184, 193, 206
Ordinality, 47, 49, 50, 51, 52, 102, 103, 104, 122, 138, 143

P

Partitioning, 151
Pattern recognition, 154, 155, 161, 162, 234, 235, 237
Pattern running, 195
Perceived numbers, 52
Plurality, 226, 227
Productive symbol mode, 42, 88
Protocounting, 109, 121, 135, 136, 137
Prototype matching, 95, 100, 136, 137, 139, 140, 141, 142, 143, 144

R

Reaction time (RT) procedure, 150
Receptive symbol mode, 42

SUBJECT INDEX

Reference memory, 181, 182, 183, 185, 195
Relative numerousness, 61
Relative numerousness judgments, 110, 135, 136, 137, 142, 143, 144, 231
Relative numerosity, 61, 62, 109, 110, 238
Relative numerosity judgment, 10, 79
Relative prototype matching, 136
Repetition recognition, 246, 247, 248, 249
Response counting, 172
Rhythm, 30, 114, 121, 156, 157, 164, 165, 201, 202, 235

S

Sequential counting, 245, 246
Sequential subitizing, 164
Serial order, 245, 246, 253, 254, 259
Signal detection, 25, 26, 27
Simultaneous counting, 19, 20, 201, 215, 216
Stable-order principle, 137, 183, 184, 193, 206, 218, 222, 245, 253
Stimulus counting, 172

Stimulus generalization, 132, 133
Subitizing, 122, 131, 134, 135, 136, 137, 139, 140, 141, 149, 152, 154, 157, 166, 194, 200, 201, 219, 232, 233, 234, 236, 237, 245
Subtraction, 213
Successive counting, 20, 21, 215, 216, 217, 245
Summation, 44, 45, 46, 91, 92, 93
Symbolic counting, 44, 45, 88

T

Tagging, 46, 47, 48, 49, 99, 137, 138, 158
Threshold procedure, 150
Timing, 9, 11, 27, 171, 174, 176, 177, 185, 186, 200, 220, 221
Transitive inference, 47, 49, 50, 51, 52, 122
Trial chunk, 191, 207

W

Working memory, 181, 182, 183, 196